Green Technology and Renewable Energy Projects

Green Technology and Renewable Energy Projects

Editor

Mostafa Ghasemi Baboli

MDPI • Basel • Beijing • Wuhan • Barcelona • Belgrade • Manchester • Tokyo • Cluj • Tianjin

Editor
Mostafa Ghasemi Baboli
Sohar University
Oman

Editorial Office
MDPI
St. Alban-Anlage 66
4052 Basel, Switzerland

This is a reprint of articles from the Special Issue published online in the open access journal *Sustainability* (ISSN 2071-1050) (available at: https://www.mdpi.com/journal/sustainability/special_issues/green_technology_renewable_energy_projects).

For citation purposes, cite each article independently as indicated on the article page online and as indicated below:

LastName, A.A.; LastName, B.B.; LastName, C.C. Article Title. *Journal Name* **Year**, *Volume Number*, Page Range.

ISBN 978-3-0365-3941-6 (Hbk)
ISBN 978-3-0365-3942-3 (PDF)

Contents

About the Editor

Mostafa Ghasemi Baboli

Dr. Mostafa was born in Iran and graduated from the Petroleum University of Technology in 2006. He obtained his PhD in Chemical Engineering from the National University of Malaysia (UKM) in 2012. In 2012, his manuscript in the Chemical Engineering Journal was selected as a top cited manuscript. Until March 2016 he served as Senior Lecturer/Research Fellow at the National University of Malaysia. In March 2016, he joined Universiti Teknologi PETRONAS (UTP). In addition to his responsibilities of teaching and research, he has been the principle investigator of an International Project with King Abdul Aziz City for Science and Technology (KACST). Dr Mostafa has supervised more than 20 theses, published over 80 journal articles and a number of chapter books. At the end of 2019, he joined the Chemical Engineering department of Sohar University as an Associate Professor. Until now, he has about 10 years of professional experience in biochemical fuel cells renewable energy and membrane technology for water and wastewater treatment. He is also a trainer for writing up scientific manuscripts.

Preface to "Green Technology and Renewable Energy Projects"

With increasing population growth, the demand for water and energy has increased significantly. The consequences from a lack of water and energy will cause poverty and many problems in various societies. Additionally, it will lead to serious environmental problems such as global warming, drought, and water scarcity. In this book, we collected some new studies on the importance of different techniques in the production of renewable energies and applying green technologies.

The editor would like to thank the authors of the different chapters of the book. The assistance and help provided by Sohar University is also highly appreciated.

Mostafa Ghasemi Baboli

Editor

sustainability

MDPI

Article

Carbon Nanotube/Pt Cathode Nanocomposite Electrode in Microbial Fuel Cells for Wastewater Treatment and Bioenergy Production

Mostafa Ghasemi [1,*], Mehdi Sedighi [2] and Yie Hua Tan [3]

1 Chemical Engineering Section, Faculty of Engineering, Sohar University, Sohar 311, Oman
2 Department of Chemical Engineering, University of Qom, Qom 3716146611, Iran; sedighi@qom.ac.ir
3 Department of Environmental Engineering, Faculty of Engineering and Science, Curtin University, Miri 98009, Sarawak, Malaysia; tanyiehua@curtin.edu.my
* Correspondence: MBaboli@su.edu.om

Abstract: In this paper, we reported the fabrication, characterization, and application of carbon nanotube (CNT)-platinum nanocomposite as a novel generation of cathode catalyst in microbial fuel cells (MFCs) for sustainable energy production and wastewater treatment. The efficiency of the carbon nanocomposites was compared by platinum (Pt), which is the most effective and common cathode catalyst. This nanocomposite is utilized to benefit from the catalytic properties of CNTs and reduce the amount of required Pt, as it is an expensive catalyst. The CNT/Pt nanocomposites were synthesized via a chemical reduction technique and the electrodes were characterized by field emission scanning electron microscopy, electronic dispersive X-Ray analysis, and transmission electron microscopy. The nanocomposites were applied as cathode catalysts in the MFC to obtain polarization curve and coulombic efficiency (CE) results. The catalytic properties of electrodes were tested by linear sweep voltammetry. The CNT/Pt at the concentration of 0.3 mg/cm^2 had the highest performance in terms of CE (47.16%), internal resistance (551 Ω), COD removal (88.9%), and power generation (143 mW/m^2). In contrast, for the electrode with 0.5 mg/L of Pt catalyst, CE, internal resistance, COD removal, and power generation were 19%, 810 Ω, 96%, and 84.1 mW/m^2, respectively. So, it has been found that carbon nanocomposite cathode electrodes had better performance for sustainable clean energy production and COD removal by MFC.

Keywords: carbon nanotube; coulombic efficiency; microbial fuel cell; nanocomposite; Pt

Citation: Ghasemi, M.; Sedighi, M.; Tan, Y.H. Carbon Nanotube/Pt Cathode Nanocomposite Electrode in Microbial Fuel Cells for Wastewater Treatment and Bioenergy Production. *Sustainability* **2021**, *13*, 8057. https://doi.org/10.3390/su13148057

Academic Editor: Farooq Sher

Received: 13 June 2021
Accepted: 17 July 2021
Published: 19 July 2021

Publisher's Note: MDPI stays neutral with regard to jurisdictional claims in published maps and institutional affiliations.

1. Introduction

Climate change and the risk of running out of non-renewable energy has made scientists think about clean alternative types of fuels [1–3]. In addition, as a consequence of population growth and the industrial revolution, the energy demand has been on an upward trend, and also a large volume of wastewater is produced [4,5]. Therefore, there is a considerable request globally for reliable, clean energy sources and wastewater treatment plants. The microbial fuel cells (MFC) are an apparatus that converts the chemical energy of biodegradable organic compounds to electricity and hydrogen, which are the clean and renewable types of energy [6,7]. This means that an MFC can utilize the organic compound of waste and produce energy, thus, treating wastewater [8–11] and producing energy simultaneously [7]. Generally, MFCs consist of two chambers, namely cathode, and anode divided by a separator or proton exchange membrane (PEM) [12,13]. The anode chamber utilizes organic matter to produce electrons, which are transferred to the cathode chamber via an external circuit. The produced protons will pass through the separator to reach the cathode chamber. Lots of efforts have been made to make the MFCs commercial [14,15]. Most of the expenses of an MFC are due to its cathode catalyst (Pt), which accelerates the oxygen reduction reaction (ORR). According to the reports, more than half of MFC price is related to Pt [16,17]. Platinum is a durable and efficient catalyst, which is adopted for MFC and makes the ORR faster. However, it is very expensive and is not applicable for use on a large scale. Therefore, researchers tried to find a catalyst other than Pt or decrease the amount of Pt in the MFC. Among all the materials, carbon nano-based materials and nanocomposites attracted lots of attention due to their unique properties. They possess a substantial specific surface area with high catalytic activity and showed promising properties in electron production and transfer [18–20]. Jafar Ali et al. [21] used nanocomposite iron sulfide wrapped with graphene oxide as the cathode catalyst. The MFC with nanocomposite cathode catalyst presented a higher removal of toxic Cr and a 4.6 higher reduction rate than blank MFC [22]. Moreover, the MFC which used nanocomposite cathode catalyst produced 3.28 times more power density. Another group of researchers [23] tried to replace the expensive cathode catalyst. They used two different multiwall carbon nanotube (CNT)-based cathode catalysts. These authors found that Co-N-CNT and Fe-N-CNT had higher ORR activity and also generated more power density than Pt [24]. In another, Pattanayak et al. [22] utilized poly (aniline co pyrrole) wrapped titanium dioxide nanocomposite for an air cathode MFC. They found that the nanocomposite catalyst produced 2.3 times higher power output than the Pt/c catalyst. In another interesting study, Yu Du et al. [25] replaced Pt as an expensive cathode catalyst with a hybrid nanocomposite catalyst. They used nitrogen-doped CNT, reduced graphene oxide nanosheet, and compared this new catalyst with Pt. These researchers could produce 1329 mW/cm^2 power density which was 1.37 times higher than Pt catalyst.

In a recent investigation, Yan Wang et al. [26] synthesized a bimetallic hybrid modified with CNT and applied it as a cathode catalyst in MFC. The newly synthesized cathode catalyst had higher ORR performance and produced 2757 mW/m^3 power density which was higher than Pt (2313 mW/m^3) catalyst.

In 2019, Majidi et al. fabricated α-MnO_2/C, and α-MnO_2/C supported on carbon Vulcan catalyst for air cathode MFC. The catalysts could produce 180 and 111 mW/m^2, respectively. However, the produced powers were not that much higher, but the catalysts were considered as economical compared to the pure Pt catalyst [27].

Sofia et al. have done a study on platinum group metal-free based catalyst in 2020. However, they reported that the power output is increased but the generated power was 67.3 mW/m^2 and 120 $mW\ m^{-2}$ after 35 days and two months respectively. The power densities were quite low and also the power output changed a lot due to the fluctuating environmental condition [28].

As different parameters influence the performance of MFC, Ghasemi et al. [24] used two methods for optimizing MFC. They tried artificial intelligence and fuzzy logic. The mentioned authors considered diverse optimized conditions and interestingly observed that each optimization technique produced a certain result.

However, there are lots of studies about synthesis and applications of new catalysts in MFCs, there are not many studies about optimization of Pt and making MFCs more economical by Pt composite catalysts. As the Pt is the most efficient and durable catalyst and also the most expensive, we tried to reduce the amount of Pt, but not eliminate it.

This research is a big step towards the commercialization of MFC because it use a smaller amount of Pt as the cathode catalyst, thus making MFC a sustainable method for use on a bigger scale for societies and industries. Until now, MFC has not been applied widely as it is not a sustainable solution for wastewater treatment and energy production. However, the scientists had lots of progress in improving the MFCs for higher power generation and wastewater treatment [4].

In the present study, CNT/Pt nanocomposite has been synthesized and characterized. Furthermore, the optimum amount of this catalyst was found for using in MFC to reach higher ORR potential, COD removal, and power generation. The self-generated nanocomposite was made and applied as a cathode catalyst in MFC and the performance of this option was compared with Pt catalyst.

2. Materials and Methods

2.1. MFC Configuration

The design of MFC was the same as in previous studies [29]. It consisted of two cylindrical chambers with a length of 14 cm and a diameter of 6.2 cm (300 cm^3 active volume) separated by Nafion 117 (China). The membrane was pretreated before use by boiling water and 3% hydrogen peroxide, or 0.5 M sulfuric acid for half an hour and kept inside deionized water. The cathode and anode electrodes had 12 cm^2 surface area [29]. The 0.5 M phosphate buffer solution was used to adjust the pH of the solution around 6.5–7. The cathode chamber consisted of 4.26 g/L Na_2PO_4, 2.5 g/L NaH_2PO_4, 0.13 g/L KCl, and 0.31 g/L NH_4Cl solution, and aeration was performed by an aquarium pump. Carbon paper (CP) was utilized as the anode electrode, in addition to Pt and CNT/Pt being employed at different concentrations as the cathode electrode. Figure 1 shows the schematic of an MFC. The protons passed through PEM and electrons were transferred through the external circuit to reach the cathode.

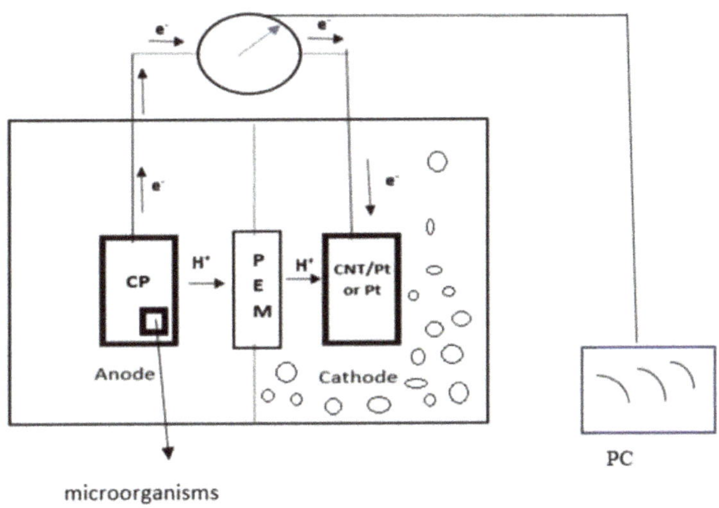

Figure 1. Schematic representation of the fabricated MFC.

For inoculation of MFC at the anode chamber, anaerobic sludge from a palm oil mill effluent treatment plant (Selangor, Malaysia) was utilized. The media comprised of 5 g glucose as the source of sugar, 0.5 g yeast extract, 4 g $NaHCO_3$, 0.2 g KCl, and 10 mL Wolfe's mineral and vitamin solution added per liter [30]. All the chemicals were provided by Merck Company (Darmstadt, Germany).

2.2. Electrode Preparation

2.2.1. Pt Electrode

For preparing Pt electrode, first, the Pt was washed with deionized water followed by filtering and drying in the oven. Next, Pt was dispersed in a Nafion solution resulting in Pt ink. The ink was dispersed on the CP electrode by a brush and was dried for 1 h at 100 °C in an oven [12].

2.2.2. CNT/Pt Nanocomposite Electrode

To produce CNT/Pt, the chemical reduction technique was used. In the first step, CNT was dispersed by ultrasonication in HNO_3 for 3 h. Afterward, the produced solution was dried, rinsed, and finally dried in the air. The resultant CNT sample was then dispersed by ultrasonication in $(CH_3)_2CO$ for 1 h. Next, the 0.075 molar H_2PtCl_6 solution was slowly added to CNT during stirring. After 24 h, the mixture was reduced by 0.1 M $NaBH_4$ as well as 1 M NaOH solution. As soon as the mixture was uniformly mixed, it was rinsed and dried at 80 °C for almost 6 h [16]. The required quantity of CNT/Pt was added to a limited proportion of C_2H_5OH, dispersed properly on a mixer, and brushed on the CP surface. Afterward, the CNT/Pt electrode was placed in the oven at 100 °C to dry.

2.3. Analysis and Calculation

The experiments were done in three months and the data were collected in three days (72 h) when the system became stable.

Formulas (1) and (2) were utilized to determine the current as well as power density:

$$I = \frac{V}{R} \tag{1}$$

$$P = R \times I^2 \tag{2}$$

where I denotes the current (amps), R is the external resistance (ohm), V refers to the voltage (V), and P represents the produced power (Watt) of the system [31,32]. The coulombic efficiency (CE) was calculated as the actual current transferred to the maximum current that is obtainable from the system. It can be calculated by the following equation [18]:

$$CE = \frac{M \int_0^t I\,dt}{F\, b\, V_{an}\, \Delta COD} \tag{3}$$

where M refers to the molecular weight of oxygen (32 g/mole), F denotes the Faraday's constant (96,485 C/mol-electrons), $b = 4$ is the number of electron exchange per mole of oxygen, ΔCOD represents the change of COD over time t, and V_{an} is the liquid volume (m^3) in the anode chamber.

For measurement of the COD, first, the samples of media were diluted 10 times and then 2 mL of the dilute was mixed uniformly with (a vial) of high-range COD reagent digestion solution. Then it was heated to 150 °C by a thermo reactor (Hach, DRB 200, Loveland, CO, USA) for 2 h, and read with a spectrophotometer (Hach, DR 2800, Loveland, CO, USA) [33].

The experiments were performed three times under the same conditions and the mean values or typical results are presented below.

2.4. Electrodes Morphology and Catalytic Activity

The morphology of the surface of the electrodes and the attached bacterial community on the anode electrode was observed by field emission scanning electronic microscopy (FESEM) (Supra 55vp-Zeiss, Oberkochen, Germany). Before the FESEM analysis, the samples were entirely coated by a thin conducting metal, such as gold. Transmission electronic microscopy (TEM) (CM-12, Philips, Eindhoven, The Netherlands) was employed to find out the dispersion of Pt nanoparticles. The Energy-Dispersive X-ray spectroscopy (EDX) (INCA, Oxford, UK) was applied to detect the percentage of Pt in the nanocomposite. The catalytic activity of the electrodes was tested by a potentiostat galvanostat (HAK-MILIK FRIM 04699A-2007, Japan) in 0.1 M H_2SO_4 with the reference electrode of Ag/AgCl, working electrode of glassy carbon and Pt as counter electrode for linear sweep voltammetry (LSV) and 50 mV/S scan rate in the range of −1 until 0.4 V [34].

3. Results and Discussion

3.1. Electrode Characterization

Figure 2a,b show the FESEM, as well as TEM images of Pt-dispersed CNTs. The uniform dispersion of Pt nanoparticles in the CNTs can be observed in the TEM image, where the dark spots indicate the presence of Pt. The functionalization of CNTs using HNO_3 increased the surface reactivity and active sites of CNTs facilitating the fine dispersion of Pt nanoparticles on the CNT surface [35]. It enhances electrode electrochemical activity due to the high catalytic surface area of CNT and the supportive and catalytic effects of Pt [36]. The EDX pattern shown in Figure 2c indicates the percentage of CNT and Pt in the nanocomposite. The Pt weight percentage to CNT ratio was found to be around 25%.

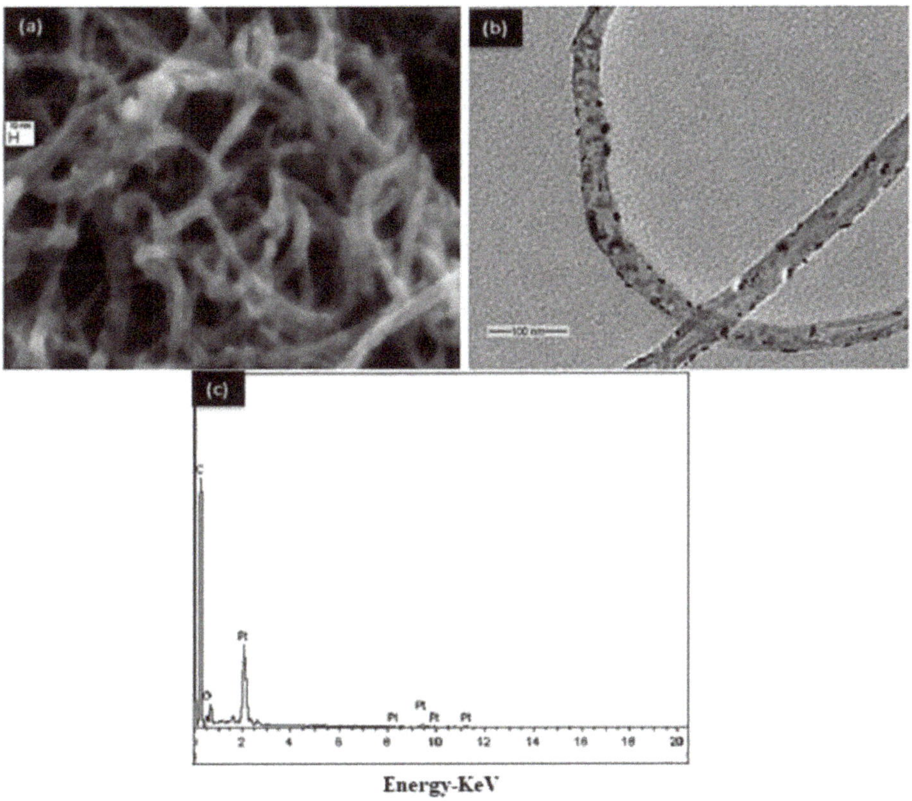

Figure 2. (**a**) FESEM, (**b**) TEM, and (**c**) EDX pattern of CNT/Pt nanocomposite.

3.2. *Attachment of Microorganisms on the Anode Electrode*

Figure 3 shows the morphology of the anode electrode and the attachment of microorganisms. It indicates clearly that various types of bacteria are attached to the electrode surface. They utilize the organic substrate of the feed, growing and producing electrons and protons which will be transferred to the cathode for electricity production [37].

Glucose was used as the substrate in the MFC. The reactions which occurred at the cathode and anode are summarized in Equations (4) and (5), respectively.

$$C_6H_{12}O_6 + 6H_2O \longrightarrow 6CO_2 + 24\,e^- + 24H^+ \quad (4)$$

$$O_2 + 24\,e^- + 24H^+ \longrightarrow 12H_2O \quad (5)$$

The oxidation of one mole of glucose in an anaerobic condition produces 24 mol electrons and protons. In our previous study.

Figure 3. Attachment of microorganisms on the electrode.

3.3. Power Density

By changing the external load on the MFC, a power density graph can be drawn as shown in Figure 4.

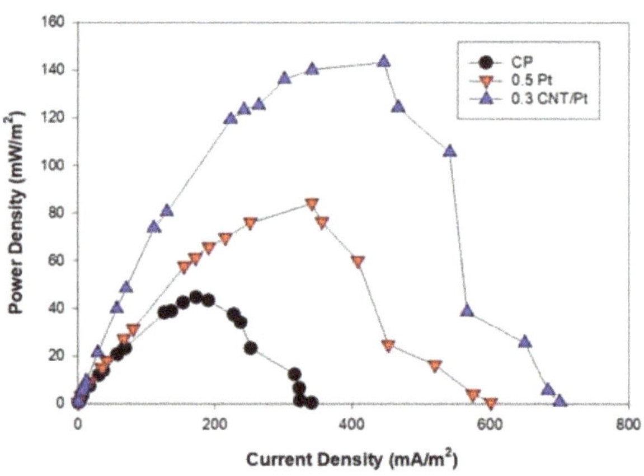

Figure 4. Power density graph of some electrodes.

The maximum power density of neat CP was 44.7 mW/m^2 at 172.66 mA/m^2. It increased by elevating the Pt amount and reached the maximum amount of 84.01 mW/m^2 (341 mA/m^2) for 0.5 mg/cm^2. However, for CNT/Pt cathode electrode, the power density augmented with rising the amount of CNT/Pt up to 0.3 mg/cm^2, in which the power density reached 143.1 mW/m^2 and remained almost constant [29].

Like previous studies [24,25], the nanocomposite cathode catalyst produced a higher power density than Pt.

The Pt nanoparticles placed on the CNT surface can absorb more hydrogen (H^+) ions due to the more open active surface, compared to Pt. The role of these particles in the dissociation of the hydrogen molecule is critical and leads to an increase in the amount of adsorbed hydrogen by Van der Waals force [12]. In addition, the power was approximately constant without augmentation in CNT/Pt of over 0.3 mg/cm^2. The latter result shows that the diffusion resistance of the cathode raised because of a depletion in the oxygen diffusion rate that offsets the elevation of the catalyst [38].

3.4. Polarization Curve

The polarization curve of different systems is demonstrated in Figure 5. Typically, the voltage will decline by an increase in current density [39].

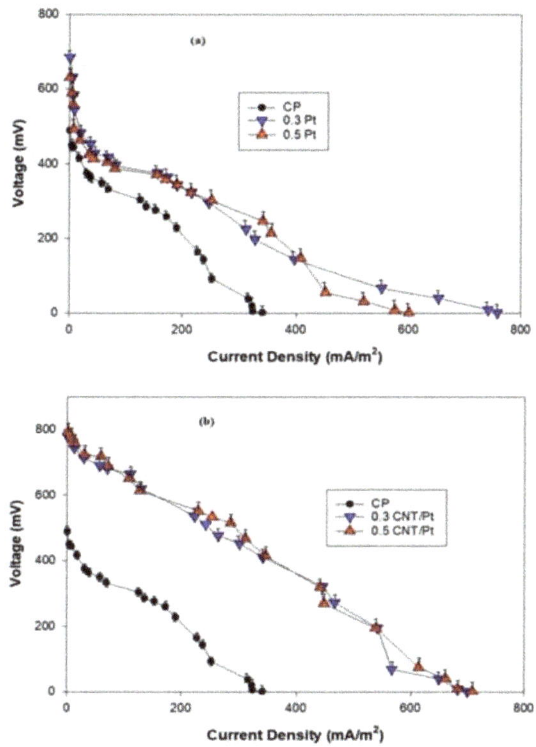

Figure 5. Effect of different kinds of electrodes on the polarization curve of fabricated MFC (**a**) Pt (**b**) CNT/Pt.

The internal resistance of the system was measured by the polarization curve and through mathematical calculations. Internal resistance is the slope of the voltage-current curve [40]. Therefore, a higher slope of the polarization curve means that the internal resistance is higher resulting in lower electricity production [41]. Table 1 indicates the internal resistance, maximum produced power, maximum current, and internal resistance at a maximum power density of different MFCs.

Table 1. Information taken from the MFCs.

Cathode Electrode	Internal Resistance (Ω)	P_{max} (mW/m^2)	I_{max} (mA/m^2) at P_{max}	OCV at SS Condition (mV)
CP	1084	44.7	172.7	493
0.1 Pt	894	53.4	159.5	587
0.2 Pt	873	77	327.1	687
0.3 Pt	812	73	246.7	693
0.4 Pt	811	75.9	251.6	673
0.5 Pt	810	84.1	341.7	672
0.1 CNT/Pt	587	73.5	221.3	746
0.2 CNT/Pt	593	125	288.7	787
0.3 CNT/Pt	551	143.1	445.8	794
0.4 CNT/Pt	564	146	312	811
0.5 CNT/Pt	573	146.8	285.6	815

Overall, the system that worked with CP as a cathode had the highest internal resistance of about 1084 Ω. The internal resistance decreased to 810 Ω by increasing the concentration of Pt to 0.5 mg/cm^2 as the lowest internal resistance among all Pt electrodes in this study. Moreover, in CNT/Pt nanocomposite, the lowest was 551 Ω for 0.3 mg/cm^2 CNT/Pt. This shows that the CNT/Pt composite had an internal resistance about 35% lower than Pt which may be attributed to the better electrical and catalytic activity, and especially higher conductivity, of the CNT/Pt nanocomposite electrode compared to Pt [42,43].

3.5. COD Removal and Coulombic Efficiency

Figure 6 demonstrates the COD removal, as well as the CE of the different systems.

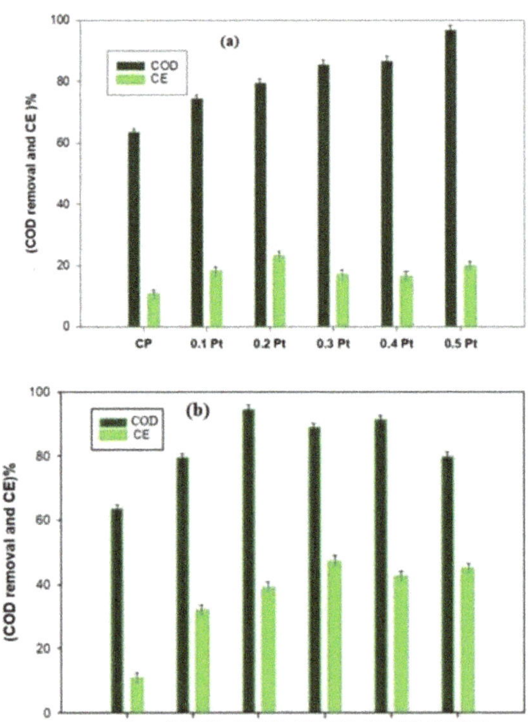

Figure 6. CE and COD removal of (**a**) Pt and (**b**) CNT/Pt electrodes.

As can be seen in Figure 6a, the highest and lowest CEs of 23% and 10.77% belonged to the CP coated by 0.2 mg/cm^2 Pt and neat CP, respectively. This means that more current is obtained from a substrate in the electrode by 0.2 mg/cm^2 Pt, in comparison with other Pt electrodes. Furthermore, the figure reveals that all electrodes coated by Pt had high COD removal of more than 70% indicating that MFCs are a proper method for COD removal [44,45]. Figure 6b shows the COD removal and CE of nanocomposite electrodes. The figure represents that the highest CE is related to the CP coated by 0.3 mg/cm^2 CNT/Pt around 47.159% in 88.9% COD removal. On the other hand, the lowest CE was 31.97% in 79.5% COD removal for 0.1 mg/cm^2 CNT/Pt. It should be noted that the COD removal for all nanocomposite electrode systems is >80%. For Pt electrodes, by increasing the concentration of Pt, COD removal and CE augment, while it did not exert the same impact on CNT/Pt. This can be observed in the internal resistance of the MFCs as well. For the Pt electrodes, by increasing the concentration of Pt, reduction occurs in internal resistance and then remains almost constant. However, it is not the same for CNT/Pt. In other words, elevation in internal resistance leads to a more complicated transfer of electrons from anode to cathode. Consequently, the power of the system for degrading organic substrates decreases resulting in diminished COD removal [46,47].

Table 2 compares the power density production and COD removal of different MFCs.

Table 2. Application of some cathode catalysts in MFC.

Cathode Catalyst	PEM	Power Density (mW/m^2)	COD Removal (%)	References
CoNiAl-LDH CoNiAl-LDH@NiCo_2O_4	Nafion N966	87.91 85.28	87.38 85.81	[48]
Pt-coated titanium	Ultrex CMI-7000)	271	-	[49]
SGO-TiO_2-PANi TiO_2-PANi Pt	Nafion 117	904.18 561.5 483.5	- - -	[50]
Pt CuZn C	Clayware based ceramic cylinder	110 75.1 19.2	90 87 77	[51]
Pt	Nafion 117 Nafion 117-SPVDF Laminated 117-SPVDF	481.55 446 413	93.97 90.27 84.15	[52]
Pt (PANi-Co-PPy)@TiO_2	Nafion 117	481 987	90.48 81.2	[22]
Pt CNT/Pt C	Nafion 117	84.1 143.1 44.7	96.8 88.9 63.5	This study

3.6. LSV Analysis

The activity of oxygen reduction on the cathode electrodes at diverse concentrations of Pt catalyst (Figure 7a) and CNT/Pt (Figure 7b) catalyst were compared together using LSV. Figure 6a indicates that the higher amount of Pt improved the ORR activity and the electrode with 0.5 mg/L Pt had the highest catalytic performance. On the other hand, for CNT/Pt, the catalytic performance augmented up to 0.3 mg/L CNT/Pt and then started to decline. This might be because the CNT weakens oxygen bonds resulting in lower activation energy. Moreover, the amount of generated current density by CNT/Pt electrodes is generally equal to or higher than the generated current by the Pt electrodes [19].

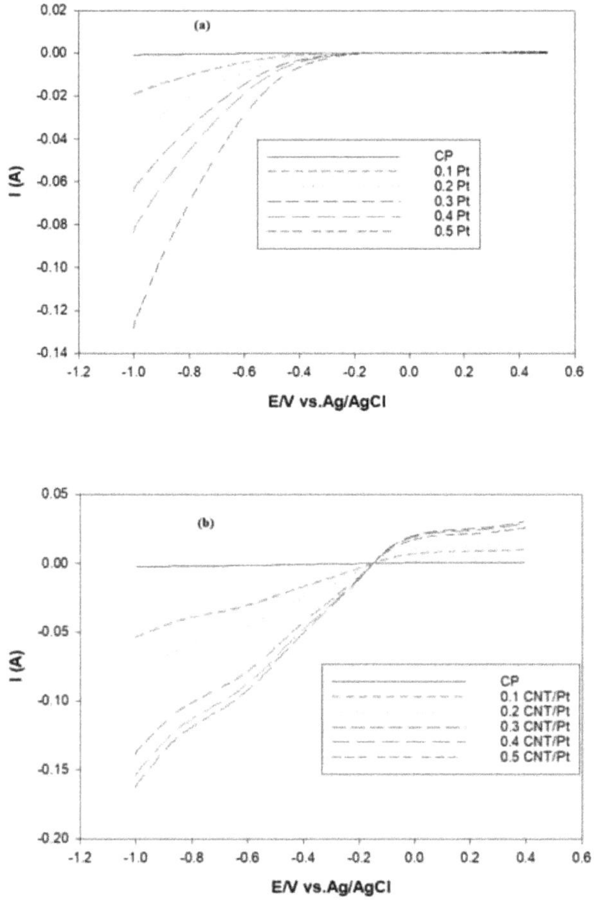

Figure 7. LSV patterns of (**a**) Pt and (**b**) CNT/Pt at different concentrations.

4. Conclusions

Sustainable energy production and wastewater treatment are some of the main parameters of industrial growth in the new era. MFC is a device that converts waste to energy. In this study, CNT/Pt nanocomposite in cathode catalysts was fabricated and applied as the cathode catalyst in MFC. It was made by drop-by-drop mixing of H_2PtCl_6 and CNT. Afterward, it was applied in the MFC for wastewater treatment and clean energy production. Our findings revealed that MFC showed better performance in terms of power generation and CE with the carbon nanocomposite cathode catalyst, compared to pure platinum.

The CNT provided more space for the attachment of Pt nanoparticles and facilitated the dissociation of protons and ORR by anticipating a higher surface area.

Furthermore, by decreasing the amount of Pt in the cathode, the capital cost of the MFC is reduced, making it more economical and applicable for societies and industries.

The nanocomposites diminished the activation energy and indicated great catalytic activity and therefore, can be a proper alternative for pure Pt. However, the optimization of the amount of nanocomposite should always be considered.

It was found that 0.3 mg/cm^2 CNT/Pt cathode catalyst produced 1.7 more power density compared to 0.5 mg/cm^2 Pt. It had also lower internal resistance (551 Ω), compared to the 0.5 mg/cm^2 Pt (810 Ω). Moreover, it had the highest CE (47.159%) and was shown to be the most active catalyst. This research opened a new window towards sustainable clean energy production and environmental remediation.

Author Contributions: Methodology, Y.H.T.; formal analysis, M.S.; resources, M.G. All authors have read and agreed to the published version of the manuscript.

Funding: This research received no external funding.

Institutional Review Board Statement: Not applicable.

Informed Consent Statement: Not applicable.

Conflicts of Interest: The authors declare no conflict of interest.

References

1. Arulmani, S.R.B.; Gnanamuthu, H.L.; Kandasamy, S.; Govindarajan, G.; Alsehli, M.; Elfasakhany, A.; Pugazhendhi, A.; Zhang, H. Sustainable bioelectricity production from *Amaranthus viridis* and *Triticum aestivum* mediated plant microbial fuel cells with efficient electrogenic bacteria selections. *Process. Biochem.* **2021**, *107*, 27–37. [CrossRef]
2. Sedighi, M.; Mohammadi, M. CO_2 hydrogenation to light olefins over Cu-CeO_2/SAPO-34 catalysts: Product distribution and optimization. *J. CO_2 Util.* **2020**, *35*, 236–244. [CrossRef]
3. Sedighi, M.; Ghasemi, M.; Mohammadi, M.; Hassan, S.H.A. A novel application of a neuro–fuzzy computational technique in modeling of thermal cracking of heavy feedstock to light olefin. *RSC Adv.* **2014**, *4*, 28390–28399. [CrossRef]
4. Sallam, E.; Khairy, H.; Elnouby, M.; Fetouh, H. Sustainable electricity production from seawater using Spirulina platensis microbial fuel cell catalyzed by silver nanoparticles-activated carbon composite prepared by a new modified photolysis method. *Biomass Bioenergy* **2021**, *148*, 106038. [CrossRef]
5. Ghasemi, M.; Mohammadi, M.; Sedighi, M. Sustainable production of light olefins from greenhouse gas CO_2 over SAPO-34 supported modified cerium oxide. *Microporous Mesoporous Mater.* **2020**, *297*, 110029. [CrossRef]
6. Obileke, K.; Onyeaka, H.; Meyer, E.L.; Nwokolo, N. Microbial fuel cells, a renewable energy technology for bio-electricity generation: A mini-review. *Electrochem. Commun.* **2021**, *125*, 107003. [CrossRef]
7. Mohammadi, M.; Sedighi, M.; Natarajan, R.; Hassan, S.H.A.; Ghasemi, M. Microbial fuel cell for oilfield produced water treatment and reuse: Modelling and process optimization. *Korean J. Chem. Eng.* **2020**, *38*, 72–80. [CrossRef]
8. Shamshiri, A.; Alimohammadi, V.; Sedighi, M.; Jabbari, E.; Mohammadi, M. Enhanced removal of phosphate and nitrate from aqueous solution using novel modified natural clinoptilolite nanoparticles: Process optimization and assessment. *Int. J. Environ. Anal. Chem.* **2020**, 1–20. [CrossRef]
9. Mohammadi, M.; Sedighi, M.; Ghasemi, M. Systematic investigation of simultaneous removal of phosphate/nitrate from water using Ag/rGO nanocomposite: Development, characterization, performance and mechanism. *Res. Chem. Intermed.* **2021**, *47*, 1377–1395. [CrossRef]
10. Sedighi, M.; Mohammadi, M. Application of green novel NiO/ZSM-5 for removal of lead and mercury ions from aqueous solution: Investigation of adsorption parameters. *J. Water Environ. Nanotechnol.* **2018**, *3*, 301–310. [CrossRef]
11. Shahveh, S.; Sedighi, M.; Mohammadi, M. A Novel Application of Combined Biological and Physical Method for Nitrate and Nitrite Removal from Water. *J. Environ. Sci. Technol.* **2020**, *22*, 183–192. [CrossRef]
12. Lin, C.-W.; Lai, C.-Y.; Liu, S.-H.; Chen, Y.-R.; Alfanti, L.K. Enhancing bioelectricity generation and removal of copper in microbial fuel cells with a laccase-catalyzed biocathode. *J. Clean. Prod.* **2021**, *298*, 126726. [CrossRef]
13. Ghasemi, M.; Ahmad, A.; Jafary, T.; Azad, A.K.; Kakooei, S.; Daud, W.R.W.; Sedighi, M. Assessment of immobilized cell reactor and microbial fuel cell for simultaneous cheese whey treatment and lactic acid/electricity production. *Int. J. Hydrogen Energy* **2017**, *42*, 9107–9115. [CrossRef]
14. Hisham, S.; Khan, F.A.; Aljlil, S.A.; Ghasemi, M. Investigating new techniques for the treatment of oil field produced water and energy production. *SN Appl. Sci.* **2019**, *1*, 646. [CrossRef]
15. Ghasemi, M.; Daud, W.R.W.; Alam, J.; Ilbeygi, H.; Sedighi, M.; Ismail, A.F.; Yazdi, M.H.; Aljlil, S.A. Treatment of two different water resources in desalination and microbial fuel cell processes by poly sulfone/Sulfonated poly ether ether ketone hybrid membrane. *Energy* **2016**, *96*, 303–313. [CrossRef]
16. Yu, B.; Tian, J.; Feng, L. Remediation of PAH polluted soils using a soil microbial fuel cell: Influence of electrode interval and role of microbial community. *J. Hazard. Mater.* **2017**, *336*, 110–118. [CrossRef]
17. Ghasemi, M.; Daud, W.R.W.; Ismail, M.; Rahimnejad, M.; Ismail, A.F.; Leong, J.X.; Miskan, M.; Ben Liew, K. Effect of pre-treatment and biofouling of proton exchange membrane on microbial fuel cell performance. *Int. J. Hydrogen Energy* **2013**, *38*, 5480–5484. [CrossRef]
18. Ghasemi, M.; Shahgaldi, S.; Ismail, M.; Kim, B.H.; Yaakob, Z.; Daud, W.R.W. Activated carbon nanofibers as an alternative cathode catalyst to platinum in a two-chamber microbial fuel cell. *Int. J. Hydrogen Energy* **2011**, *36*, 13746–13752. [CrossRef]
19. Ghasemi, M.; Daud, W.R.W.; Hassan, S.H.; Oh, S.-E.; Ismail, M.; Rahimnejad, M.; Jahim, J.M. Nano-structured carbon as electrode material in microbial fuel cells: A comprehensive review. *J. Alloys Compd.* **2013**, *580*, 245–255. [CrossRef]

20. Zhang, Q.; Liu, L. A microbial fuel cell system with manganese dioxide/titanium dioxide/graphitic carbon nitride coated granular activated carbon cathode successfully treated organic acids industrial wastewater with residual nitric acid. *Bioresour. Technol.* **2020**, *304*, 122992. [CrossRef]
21. Ali, J.; Wang, L.; Waseem, H.; Djellabi, R.; Oladoja, N.; Pan, G. FeS@rGO nanocomposites as electrocatalysts for enhanced chromium removal and clean energy generation by microbial fuel cell. *Chem. Eng. J.* **2020**, *384*, 123335. [CrossRef]
22. Pattanayak, P.; Papiya, F.; Kumar, V.; Singh, A.; Kundu, P.P.; Pattanayak, P.; Papiya, F.; Kumar, V.; Singh, A.; Kundu, P.P. Performance evaluation of poly(aniline-co-pyrrole) wrapped titanium dioxide nanocomposite as an air-cathode catalyst material for microbial fuel cell. *Mater. Sci. Eng. C* **2021**, *118*, 111492. [CrossRef] [PubMed]
23. Türk, K.; Kruusenberg, I.; Kibena-Põldsepp, E.; Bhowmick, G.D.; Kook, M.; Tammeveski, K.; Matisen, L.; Merisalu, M.; Sammelselg, V.; Ghangrekar, M.; et al. Novel multi walled carbon nanotube based nitrogen impregnated Co and Fe cathode catalysts for improved microbial fuel cell performance. *Int. J. Hydrogen Energy* **2018**, *43*, 23027–23035. [CrossRef]
24. Ghasemi, M.; Nassef, A.M.; Al-Dhaifallah, M.; Rezk, H. Performance improvement of microbial fuel cell through artificial intelligence. *Int. J. Energy Res.* **2020**, *45*, 342–354. [CrossRef]
25. Du, Y.; Ma, F.-X.; Xu, C.-Y.; Yu, J.; Li, D.; Feng, Y.; Zhen, L. Nitrogen-doped carbon nanotubes/reduced graphene oxide nanosheet hybrids towards enhanced cathodic oxygen reduction and power generation of microbial fuel cells. *Nano Energy* **2019**, *61*, 533–539. [CrossRef]
26. Wang, Y.; Zhong, K.; Li, H.; Dai, Y.; Zhang, H.; Zuo, J.; Yan, J.; Xiao, T.; Liu, X.; Lu, Y.; et al. Bimetallic hybrids modified with carbon nanotubes as cathode catalysts for microbial fuel cell: Effective oxygen reduction catalysis and inhibition of biofilm formation. *J. Power Sources* **2021**, *485*, 229273. [CrossRef]
27. Majidi, M.R.; Farahani, F.S.; Hosseini, M.G.; Ahadzadeh, I. Low-cost nanowired α-MnO_2/C as an ORR catalyst in air-cathode microbial fuel cell. *Bioelectrochemistry* **2019**, *125*, 38–45. [CrossRef]
28. Babanova, S.; Santoro, C.; Jones, J.; Phan, T.; Serov, A.; Atanassov, P.; Bretschger, O.; Babanova, S.; Santoro, C.; Jones, J.; et al. Practical demonstration of applicability and efficiency of platinum group metal-free based catalysts in microbial fuel cells for wastewater treatment. *J. Power Sources* **2021**, *491*, 229582. [CrossRef]
29. Ghasemi, M.; Halakoo, E.; Sedighi, M.; Alam, J.; Sadeqzadeh, M.; Ghasemi, M.; Halakoo, E.; Sedighi, M.; Alam, J.; Sadeqzadeh, M. Performance Comparison of Three Common Proton Exchange Membranes for Sustainable Bioenergy Production in Microbial Fuel Cell. *Procedia CIRP* **2015**, *26*, 162–166. [CrossRef]
30. Khan, F.A.; Hisham, S.; Ghasemi, M. Oil field produced water recovery and boosting the quality for using in membrane less fuel cell. *SN Appl. Sci.* **2019**, *1*, 510. [CrossRef]
31. Jafary, T.; Aljlil, S.A.; Alam, J.; Ghasemi, M. Effect of the Membrane Type and Resistance Load on the Performance of the Microbial Fuel Cell: A Step ahead of Microbial Desalination Cell Establishment. *J. Jpn. Inst. Energy* **2017**, *96*, 346–351. [CrossRef]
32. Shamshirgaran, S.; Al-Kayiem, H.; Sharma, K.; Ghasemi, M. State of the Art of Techno-Economics of Nanofluid-Laden Flat-Plate Solar Collectors for Sustainable Accomplishment. *Sustainability* **2020**, *12*, 9119. [CrossRef]
33. Sedighi, M.; Aljlil, S.A.; Alsubei, M.D.; Ghasemi, M.; Mohammadi, M. Performance optimisation of microbial fuel cell for wastewater treatment and sustainable clean energy generation using response surface methodology. *Alex. Eng. J.* **2018**, *57*, 4243–4253. [CrossRef]
34. Ghasemi, M.; Daud, W.R.W.; Rahimnejad, M.; Rezayi, M.; Fatemi, A.; Jafari, Y.; Somalu, M.R.; Manzour, A. Copper-phthalocyanine and nickel nanoparticles as novel cathode catalysts in microbial fuel cells. *Int. J. Hydrogen Energy* **2013**, *38*, 9533–9540. [CrossRef]
35. Bharti, A.; Cheruvally, G.; Muliankeezhu, S. Microwave assisted, facile synthesis of Pt/CNT catalyst for proton exchange membrane fuel cell application. *Int. J. Hydrogen Energy* **2017**, *42*, 11622–11631. [CrossRef]
36. Zhao, N.; Ma, Z.; Song, H.; Xie, Y.; Zhang, M. Enhancement of bioelectricity generation by synergistic modification of vertical carbon nanotubes/polypyrrole for the carbon fibers anode in microbial fuel cell. *Electrochim. Acta* **2019**, *296*, 69–74. [CrossRef]
37. Ben Liew, K.; Daud, W.R.W.; Ghasemi, M.; Leong, J.X.; Lim, S.S.; Ismail, M. Non-Pt catalyst as oxygen reduction reaction in microbial fuel cells: A review. *Int. J. Hydrogen Energy* **2014**, *39*, 4870–4883. [CrossRef]
38. Wang, X.; Yuan, C.; Shao, C.; Zhuang, S.; Ye, J.; Li, B. Enhancing oxygen reduction reaction by using metal-free nitrogen-doped carbon black as cathode catalysts in microbial fuel cells treating wastewater. *Environ. Res.* **2020**, *182*, 109011. [CrossRef]
39. Negassa, L.W.; Mohiuddin, M.; Tiruye, G.A. Treatment of brewery industrial wastewater and generation of sustainable bioelectricity by microbial fuel cell inoculated with locally isolated microorganisms. *J. Water Process. Eng.* **2021**, *41*, 102018. [CrossRef]
40. Elshobary, M.E.; Zabed, H.M.; Yun, J.; Zhang, G.; Qi, X.; Elshobary, M.E.; Zabed, H.M.; Yun, J.; Zhang, G.; Qi, X. Recent insights into microalgae-assisted microbial fuel cells for generating sustainable bioelectricity. *Int. J. Hydrogen Energy* **2021**, *46*, 3135–3159. [CrossRef]
41. Rossi, R.; Logan, B.E. Unraveling the contributions of internal resistance components in two-chamber microbial fuel cells using the electrode potential slope analysis. *Electrochim. Acta* **2020**, *348*, 136291. [CrossRef]
42. Rahimnejad, M.; Ghasemi, M.; Najafpour, G.; Ismail, M.; Mohammad, A.W.; Ghoreyshi, A.; Hassan, S.H. Synthesis, characterization and application studies of self-made Fe_3O_4/PES nanocomposite membranes in microbial fuel cell. *Electrochim. Acta* **2012**, *85*, 700–706. [CrossRef]
43. Parkhey, P.; Sahu, R. Microfluidic microbial fuel cells: Recent advancements and future prospects. *Int. J. Hydrogen Energy* **2021**, *46*, 3105–3123. [CrossRef]

44. Xin, S.; Shen, J.; Liu, G.; Chen, Q.; Xiao, Z.; Zhang, G.; Xin, Y. Electricity generation and microbial community of single-chamber microbial fuel cells in response to Cu_2O nanoparticles/reduced graphene oxide as cathode catalyst. *Chem. Eng. J.* **2020**, *380*, 122446. [CrossRef]
45. Ullah, Z.; Zeshan, S. Effect of substrate type and concentration on the performance of a double chamber microbial fuel cell. *Water Sci. Technol.* **2020**, *81*, 1336–1344. [CrossRef]
46. Li, S.; Ho, S.-H.; Hua, T.; Zhou, Q.; Li, F.; Tang, J. Sustainable biochar as electrocatalysts for the oxygen reduction reaction in microbial fuel cells. *Green Energy Environ.* **2020**. [CrossRef]
47. Gadkari, S.; Gu, S.; Sadhukhan, J. Two-dimensional mathematical model of an air-cathode microbial fuel cell with graphite fiber brush anode. *J. Power Sources* **2019**, *441*, 227145. [CrossRef]
48. Khajeh, R.T.; Aber, S.; Zarei, M. Comparison of $NiCo_2O_4$, CoNiAl-LDH, and CoNiAl-LDH@$NiCo_2O_4$ performances as ORR catalysts in MFC cathode. *Renew. Energy* **2020**, *154*, 1263–1271. [CrossRef]
49. Taskan, E.; Hasar, H.; Taskan, E.; Hasar, H. Comprehensive Comparison of a New Tin-Coated Copper Mesh and a Graphite Plate Electrode as an Anode Material in Microbial Fuel Cell. *Appl. Biochem. Biotechnol.* **2015**, *175*, 2300–2308. [CrossRef]
50. Papiya, F.; Pattanayak, P.; Kumar, V.; Das, S.; Kundu, P.P.; Papiya, F.; Pattanayak, P.; Kumar, V.; Das, S.; Kundu, P.P. Sulfonated graphene oxide and titanium dioxide coated with nanostructured polyaniline nanocomposites as an efficient cathode catalyst in microbial fuel cells. *Mater. Sci. Eng. C* **2020**, *108*, 110498. [CrossRef]
51. Das, I.; Das, S.; Ghangrekar, M. Application of bimetallic low-cost CuZn as oxygen reduction cathode catalyst in lab-scale and field-scale microbial fuel cell. *Chem. Phys. Lett.* **2020**, *751*, 137536. [CrossRef]
52. Kumar, V.; Kumar, P.; Nandy, A.; Kundu, P. Fabrication of laminated and coated Nafion 117 membranes for reduced mass transfer in microbial fuel cells. *RSC Adv.* **2016**, *6*, 21526–21534. [CrossRef]

sustainability

MDPI

Article

The AgTech Startup Perspective to Farmers Ex Ante Acceptance Process of Autonomous Field Robots

Friedrich Rübcke von Veltheim * and Heinke Heise

Department of Agricultural Economics and Rural Development, Georg-August-University of Göttingen, Platz der Göttinger Sieben 5, 37073 Göttingen, Germany; heinke.heise@agr.uni-goettingen.de
* Correspondence: veltheim@uni-goettingen.de; Tel.: +49-551-39-24264

Received: 31 October 2020; Accepted: 14 December 2020; Published: 17 December 2020

Abstract: Autonomous vehicles not only provide a new impetus in the development of car models in the automotive industry—even in agriculture there has recently been talk of autonomous field robots (AFR). Great expectations are placed on these digital assistants from a wide variety of perspectives. However, it is still unclear whether they will make the transition from market niches to broad-based distribution. Apart from various factors, this depends on user acceptance of this new technology expected by the innovators, since this is likely to be essential for the further development of AFR. For this purpose, the ex ante user acceptance of farmers from the perspective of various AgTech startups with AFR involvement in Europe was investigated in this exploratory and qualitative study. The Technology Acceptance Model (TAM) served as the basis for the developed interview guideline. In summary, the results confirm that a variety of factors potentially influence farmer acceptance and AFR diffusion from the perspective of AgTech startups, with perceived usefulness being considered the main motivation for using AFR. The interviewed experts believe that AFR will initially be used in crops that have relatively high costs for crop protection treatments before becoming economically attractive for other crops. The basic prerequisite for a successful market launch is an adjustment of the legal framework, which sets standards in relation to AFR and thus, provides security in the production process. The results could support political decision-makers in dealing with this new technology and AFR manufacturers in the promotion of AFR.

Keywords: acceptance; AgTech startups; autonomous; robot; TAM

1. Introduction

The megatrend of digitalization has not stopped at the agricultural sector. Rapidly developing technologies and infrastructures are opening up new perspectives that could even lead to a rethinking of entire farming systems [1]. For example, tractors as the main work equipment for arable farmers are slowly but surely being joined in their development by highly accurate, sensor-based, and intelligently linked field robots that can perform certain tasks autonomously, much more precisely, and much more sustainably either alone or in swarms with other units. The development of such autonomous field robots (AFR), which are becoming increasingly economically attractive, is mainly being driven by the growing demand for qualified workers in agriculture and the public debate about the impact of farming practices on the environment (e.g., the use of pesticides, over-fertilization, and soil compaction) and thus, addresses all three pillars of sustainability (economic, environmental, and social) [2]. Social majorities can generate political pressure, which can lead to legal changes in pesticide and fertilizer application and force farmers to adapt their farming practices. AFR, some of which are still at the prototype stage, can perform specific tasks without an operator, apply pesticides on a plant-by-plant basis, or control weeds mechanically, which is why AFR could become more important in the future [3]. With its autonomous operation, these machines reduce the workload of farmers and protect them from unnecessary contact with harmful chemicals. Due to AFR's better scalability, smallholder farms, which have often been considered uneconomical, could become more economically attractive by adopting such robots, which could lead to a rethink in process in the agricultural sector, away from the motto: "bigger is better" [4].

However, the introduction of autonomous technologies in agriculture is also accompanied by some concerns. For instance, although the driving of farm machinery would be abolished, new tasks would be added, such as AFR's monitoring and programming, for which farmers may lack qualified staff [3]. In addition, Devitt [5] fears that farmers may not be able to trust unmanned robots working out of sight and that by handing over tasks to artificial intelligence (AI), they would lose agricultural knowhow in the long run and thus, suffer an even greater loss of social recognition. A similar conclusion was reached by Kester et al. [6], who, using the example of special crops, revealed a conflict between the farmers' support of efficiently operating AFR and a lack of confidence in such technologies.

Despite the concerns expressed above, there is broad agreement that AFR will claim its place in modern agriculture in the future, but it remains unclear when and to what extent this will happen. For instance, the King [4] argues that the initial reluctance of established agricultural machinery manufacturers to develop AFR is due to the fact that their existing business models are being compromised. In addition, the risk of damaging reputation by a possible malfunction of the first AFR is much higher for established agricultural machinery manufacturers than for largely unknown AgTech (AgTech (Agricultural Technology) is used in this article as a generic term for all technical innovations affecting the data-driven, networked, digital agriculture of the future) startups. However, AFR also offer a whole range of new business models for agricultural machinery manufacturers, such as on-demand supply of AFR to farmers in return for a rental fee or as a cloud-based pay-per-use model [7].

Another opportunity can be seen in the ongoing development of agricultural engineering through a disaggregated approach in the form of several small robot units where large machines hit the legal size and weight limits [8]. Since digital technologies such as AFR collect and analyze large amounts of data in order to fully unfold their potential, not only AgTech startups but also large IT companies such as IBM and Google are competing with established agricultural machinery manufacturers. As these new players will claim their share of the market, it is all the more important for the current market leaders to assess whether there is indeed a future market for AFR and, if so, to what extent [9]. Therefore, when estimating the factors influencing the ex ante acceptance of AFR by farmers, the perspective of various AgTech startups as new market entrants appears to be highly interesting.

However, there have only been a few studies on the technology acceptance process in relation to AFR in agriculture so far. Redhead et al. [3] attribute this to the fact that this is a new, potentially disruptive technology and that not much is known about how AFR could be integrated into current farming practices. Therefore, they conducted contextual interviews with nine farmers from large-scale farms in Queensland (Australia). The farmers interviewed were particularly interested in the reduction of weed control costs and the time saved by the absence of an operator when using AFR. However, they also feared the reliability of the small robot units in an uncontrolled environment, the time required for monitoring, the complexity of AFR as well as the lack of required infrastructure in terms of high-speed mobile data networks. Overall, however, they were enthusiastic about a possible adoption of AFR, which was particularly true for farmers who were already using precision agriculture (PA) technologies on their farms. This can be confirmed by the findings of Salimi et al. [10], according to which farmers are more likely to adopt automated technologies the more they understand their usefulness. Therefore, the authors [10] surveyed 378 people with agricultural backgrounds in Iran in 2020 about factors influencing the adoption of agricultural automation. In 2018, Rial-Lovera [11] surveyed 14 agricultural stakeholders in California (USA) on their AFR acceptance behavior and revealed labor shortages and rising labor costs, as well as a lack of awareness of the potential benefits of AFR and the lack of compatibility between agricultural equipment as the main reasons for an ex ante acceptance of AFR by farmers. Using a wider definition of the term, Caffaro and Cavallo [12] conducted a survey on Smart Farming Technologies (STF) at a northern Italian agricultural fair and found that the sociodemographic characteristics of farmers influence their adoption behavior of STF. Overall, it can be concluded that there is a research deficit in the investigation of the technology acceptance process and diffusion of AFR in agriculture, as most studies focus only on the effect of certain facets and tend to ignore the multidimensionality of the acceptance process, leading to imperfect understanding of the underlying complexity of acceptance factors and diffusion barriers [13,14].

One group whose perspective appears particularly interesting in this context is, in addition to farmers (as buyers), the AgTech startups themselves (as suppliers). Following Roger's [15] theory of the diffusion of innovations, the identification and successful targeting of the right group of farmers should be a top priority for suppliers in order to invest their resources into attracting the right people. In addition, the supplier's knowledge of existing drivers and barriers to the diffusion of AFR as an essentially affected party is usually higher than among other stakeholders. AFR are currently in an early market introduction phase, in which it will be decided whether the "critical mass" [15] of adoptions will be reached and the technology will be successfully established to a broader market or not.

The aim of this paper is therefore to examine first the barriers to the diffusion of AFR by providing insights into the ex ante acceptance process of AFR. To this end, the present study examines various acceptance factors influencing the technology's diffusion from the perspective of AFR-developing AgTech startups. Since it is especially the user acceptance expected and already experienced by the suppliers that is essential for the further development and implementation of AFR, AgTech startups have collected data from farmers themselves. By interviewing the startups, it is possible to obtain aggregated knowledge about the acceptance of many farmers. Therefore, the data have been gathered through qualitative expert interviews at the International Forum of Agricultural Robotics (FIRA) in Toulouse (France) in December 2019.

2. Materials and Methods

Due the relatively small number of AgTech startups with AFR involvement in Europe and a lack of empirical research on the acceptance of AFR, a qualitative approach for data collection and analysis was chosen: expert interviews using a semi-standardized guideline to specify the wording and order of the questions, thus ensuring a uniform procedure [16]. Besides the fact that a quantitative survey was not suitable given the relatively small statistical population, expert interviews offer several advantages, since answering open questions requires higher cognitive effort from the respondents than answering closed questions and can therefore reveal more in-depth information [17]. In order to collect the data, we conducted interviews with experts from ten different AgTech startups specialized in AFR at the International Forum for Agricultural Robotics (FIRA) in Toulouse (France) in December 2019. The open and neutral formulated questions were split into six thematic blocks: drivers and barriers, economics, environmental impacts, legal constraints, socioeconomic impacts, and technology. Following a brief introduction to the topic, the experts were interviewed in person, leaving them free to add questions or topics not covered in the interview guide. The results were extracted using a qualitative content analysis according to [16,18], which is described in more detail in Section 2.2. For this purpose, individual statements from the interviews are assigned to previously developed, theory-based categories (described as factors in Section 2.1.) after summarizing and aggregating the information.

2.1. Study Design

This 14 question-long guide (see Appendix A) was based on the factors influencing the ex ante user acceptance of new technologies based on the Technology Acceptance Model (TAM) [19]. These factors cover perceived usefulness, perceived ease of use, and external factors. Since the subject of this study is the future use of AFR, it can be assumed that farmers have had little or no experience in dealing with this new technology, which is why the factors of attitude toward using and actual system use have been excluded from this analysis. In addition, Chuttur [20] was able to demonstrate a direct influence of perceived usefulness and perceived ease of use on the behavioral intention. According to [19], the perceived usefulness is the subjective probability with which a potential user will adopt a certain technology (in this case AFR) to improve his work performance in an organizational context. In contrast, the perceived ease of use describes the level of simplicity of a technology expected by a potential user. Both perceived usefulness and ease of use can be influenced by various external factors [10,21–23]. In order to address the diversity of perceived usefulness, this study explicitly addresses the underlying facets of an economic benefit in terms of, for example, higher harvest expectations, (labor-) cost savings or an increase in economic efficiency, and the facets of an environmental benefit. The effect of perceived ease of use on perceived usefulness is assumed to be positive, since a higher degree of simplicity of a technology is beneficial to its usefulness [10,11,23,24]. The following external factors have been added based on the available literature on the topic: compatibility, farm manager characteristics, information, legal framework, social influence, and workforce availability [3,5,6,10–12,14,23,25,26]. The external factor compatibility takes findings from diffusion theory into account, which has shown that the technical design strongly influences user acceptance [25]. Compatibility means the possibility of combining AFR with existing technology and thus, successfully integrating it into working practice. Therefore, we expect an effect on the perceived usefulness and the perceived ease of use. As another external factor, we have introduced the farm manager's characteristics in order to meet socioeconomic aspects such as age, educational level, and risk aversion [14]. Following an approach from PA research, the external factor information represents the data collected by AFR and its strategic operational use as well as associated data protection and data autonomy, for which we assume a direct influence on the perceived usefulness. The sociopolitical discussion on chemical plant protection products in Europe is leading to increasing legal restrictions on the use of pesticides in agriculture (e.g., Germany: [27]). In addition, the use of autonomous technologies on public roads is still not regulated in a way that is appropriate to the present time, which is why we assume that the legal framework has an influence on perceived usefulness [28]. We therefore defined the external factor legal framework as the degree to which a farmer believes that organizational and legal infrastructures exist to support the use of AFR. Social influence was taken into account as a further external factor to counteract the limitation of the TAM to ignore the social environment. The fact that the social environment has an influence on the acceptance of technology has been proven several times with the follow-up models TAM2 and the unified theory of acceptance and use of technology (UTAUT). Furthermore, farmers are influenced in their strategic decisions by their social environment [29], which is why we assume a direct relationship between social influence and perceived usefulness. The availability of workforce has been repeatedly cited in the screened literature as a driver for the development and adoption of AFR, which is why a direct influence on the behavioral intention to use AFR is assumed from this fifth and last external factor in this study [2,30]. The assumptions based on the factors mentioned above are shown in the following figure (Figure 1).

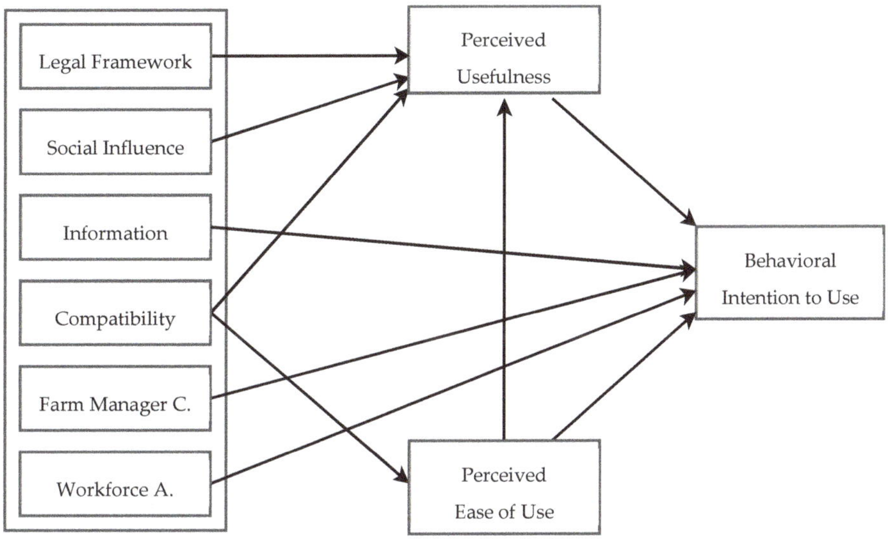

Figure 1. Modified TAM in the context of autonomous field robots.

2.2. Data Analysis

A qualitative content analysis by extraction according to [17,18] was used to analyze the data. This was performed in three steps: transcription and data exploration, identification of statements related to the acceptance factors, and summarization and interpretation of the extracted information.

The first step was to transcribe the recorded interviews in full length. The resulting transcripts were made anonymous using the method of factual anonymization. In this process, the extracted data are anonymized to such an extent that links to the individual experts can only be decrypted with a disproportionately high effort [31]. As a next step, the information relevant to the research objective of this study was extracted from the transcripts and assigned to the TAM factors described above (external factors, perceived usefulness, and perceived ease of use) that served as categories as defined by [17]. The differentiation of the factors from each other was based on their definitions, anchor examples, and coding rules (see Table A1 in the Appendix A). Since there are hardly any studies available for the research object in an agricultural context, the TAM factors were kept open in order to allow the integration of new factors which did not occur in the reviewed literature. Thus, all relevant information of the raw material could be examined under consideration of the interrelations between the factors influencing the acceptance of AFR. The methodical procedure was supported by the software "f4-analysis".

2.3. Sample Description

The ten selected AgTech startups surveyed for this study are listed in the table below (Table 1). Among these are some of the most promising competitors in this emerging industry, such as the current market leader Naïo Technologies, but also two EU-funded projects aiming to develop a market-ready product. The total number of potential AFR AgTech startups in Europe is probably in the lower two-digit range but is currently difficult to overview. Thus, the experts interviewed are very likely to cover a representative part of the total population of AgTech startups with AFR involvement in Europe.

Table 1. Sample description.

Company	Incorporation	Headquarter	AFR Involvement
Agrointelli	2015	Aarhus, Midtjylland (Denmark)	multiple task robot
AgrOnov	2015	Bretenière, Bourgogne-Franche-Comté (France)	AFR consulting
Deepfield Robotics	2008	Ludwigsburg, Baden-Württemberg (Germany)	multiple task robot
Ecorobotix	2011	Yverdon-les-Bains, Waadt (Switzerland)	solar-powered chemical weed control robot
K.U.L.T.	2012	Vaihingen an der Enz, Baden-Württemberg (Germany)	autonomous mechanical weed control
Naïo Technologies	2011	Escalquens, L'Occitanie (France)	weed control robots
Robotics for Microfarms (ROMI)	2017	Barcelona, Catalonia (Spain)	multiple task robot
SITIA	1986	Bouguenais, Pays de la Loire (France)	multiple task hybrid robot
VineScout	2016	Valencia, Valencia (Spain)	monitoring robot (vineyard)
VitiBot	2016	Reims, Grand Est (France)	multiple task robot (vineyard)

Agrointelli was founded in 2015 in Denmark, aiming to offer a fully automated farming system for arable farming. As a first step, they developed "Robotti", a versatile and autonomous multiple task robot. AgrOnov, founded in 2015 as a non-profit organization, offers a network to accompany the development of AgTech startups and to promote information and knowledge about their products. The Robert Bosch startup Deepfield Robotics emerged from a research project in 2008 with the publicly funded project "BoniRob", an autonomous robot that can, among other things, measure soil quality and remove weeds. Deepfield Robotics has recently changed its name to "farming revolution" and shifted its focus to mechanical weed control by autonomous robot platforms [32]. Ecorobotix has been developing solar-powered AFR since 2011, which control weeds with high precision using very small amounts of chemicals. The chemical company BASF is one of its investors [33]. K.U.L.T. provides innovative weeding technology for robots and has its roots back in the 1980s. After being temporarily dissolved, they have been newly founded in 2012 under the name "K.U.L.T. Kress Umweltschonende Landtechnik GmbH" and are known for their patented finger weeder. Naïo Technologies is one of the AFR pioneers, founded in 2011. They offer a variety of different AFR for weed control. To date, they have already sold nearly 150 AFR [34]. Robotics for Microfarms (ROMI) is an EU-funded project with a duration of 5 years (2017–2022) and a budget of about EUR 4 million. Their aim is to develop open and lightweight robotic platforms for microfarms, helping the farmers to reduce weeds and to monitor their crops [35]. SITIA is an industrial small- and medium-sized enterprise (SME). In 2019, SITIA launched a hybrid autonomous tractor called "TREKTOR". VineScout has also emerged from an EU-funded project (with a budget of around EUR 2 million), in which a monitoring system (decision support system) embedded in a small and cost-efficient vineyard robot is to be developed to market maturity between 2016 and 2020 [36]. Vitibot was founded in 2016. They developed the AFR "Bakus", a fully electric and autonomous straddle tractor designed to work in the vineyards.

All respondents are directly or indirectly involved in the development process of AFR and were represented as exhibitors at the International Forum of Agricultural Robotics (FIRA) in Toulouse from 8 to 10 December 2019, where the personal interviews took place.

3. Results

3.1. Perceived Usefulness

All experts agree that AFR must be economically attractive, otherwise one can only: *"(...) touch a little, little, little part of the farmers—the technophile." (The extracted statements are expressed by direct quotations, where (...) indicates that text has been skipped, while [] indicates that text has been added. This was done for practical reasons.)* (E2). For the majority of the respondents (E2, E3, E4, E5, E6, E8, E10), economic attractiveness is even the most important acceptance factor, because: *"(...) a farmer is also an entrepreneur, (...) profitability is the ultimate exclusion criterion."* (E4). Consequently, the adoption of AFR by farmers is seen as particularly important in areas where the costs of e.g., weed control (E3), workforce (E4, E5, E7), or time (E9) are relatively high. For expert 8, the price of the robot plays a less important role, since: *"(...) the return of investment is probably more important than the cost of the machine and the cost of having the machine or renting the machine just to make sure that in a short period of time you will get back the money."* (E8). Expert 9 points out that AFR are expensive to purchase but cheaper to maintain than tractors.

Most of the experts interviewed (E1, E2, E3, E6, E7, E9) also ascribe relevance to the ecological benefits of AFR for the acceptance process among farmers, as another sub-item of perceived usefulness, to the extent that they are forced to deal with resource-saving technologies due to increasing legal restrictions in plant protection. However, one expert sees in this sociopolitically driven trend towards a more sustainable agriculture also an opportunity for the development of new technologies: *"(...) the current socio-political discussion about reducing pesticides. At the moment, this is a door opener for technologies (...). That's already a big lever at the moment."* (E3). With regard to the weighting of this influence, opinions differ. Experts 2, 4, 8 and 9 agree that the ecological benefits are a good sales argument, but not the major aspect. Three other experts differentiate that this aspect is more important for organic farmers than for conventional farmers (E3, E4, E7). For two experts (E5, E10), the ecological advantages are even the most important influencing factor, whereas expert 7 believes that AFR remains a tool and that it always depends on how it is used: *"I mean all those robots, it's still tools. So, depends on how you use it. If the idea is trying to reduce the chemicals, robots can be a good tool, (...) if you want to use it in a different way [e.g.,] doing the spraying with more aggressive chemicals, because you don't have the people inside. (...) So, I think it is not the question of the tool, it's what to do with this."* (E7). The argument of battery-powered AFR concepts is also seen as controversial. For experts 2 and 5, electric engines have an acceptability enhancing effect, as they are more environmentally friendly and quieter than previous diesel engines. Expert 7 points out that the disposal of batteries is a problem that is often ignored, and for which one has to find a sweet spot. For expert 4, the ecological benefit is: *"(...) marketing. Marketing and a big lie, because robots have batteries depending on how they drive. An electric car is not more ecological than when I drive my 20-year-old diesel."* (E4).

According to most experts (E2, E3, E4, E5, E6, E8, E10), it is very important for AFR to operate reliably in order to be accepted. For about half of them (E5, E6, E8, E10), safety and reliability are even the second most important acceptance factors after economic efficiency. Experts 2, 3, 6, 7, and 8 see many teething problems with this new technology that need to be eliminated in order to increase its acceptance. Therefore, expert 7 believes that it is important that: *"(...) they [farmers] just want to see it working [in the field] and for what I understand they have enough need of this kind of tools to be able to accept"* (E7). Two of the respondents (E2, E6) add that farmers want to see that they remain in control and are not completely replaced by AFR. Expert 7 counters that there are still enough tasks that humans can do better than robots and that AFR cannot replace the farmer but only support him.

With regard to perceived usefulness, the relationships assumed at the beginning of the study (Figure 1) are reflected in many ways in the experts' statements. For example, it was emphasized several times that both the ecological and economic benefits of AFR have a direct influence on the behavioral intention to use. In addition, two new relations have been identified. First, the farm structure, as a new external factor, has a direct influence on perceived usefulness as it has a higher value for organic farms. Second, the ecological benefits of AFR appear to be increasing in value as a result of changes in the legal framework in terms of an ongoing ban on pesticides. Thus, the legal framework also has a direct influence on perceived usefulness (see Figure 2).

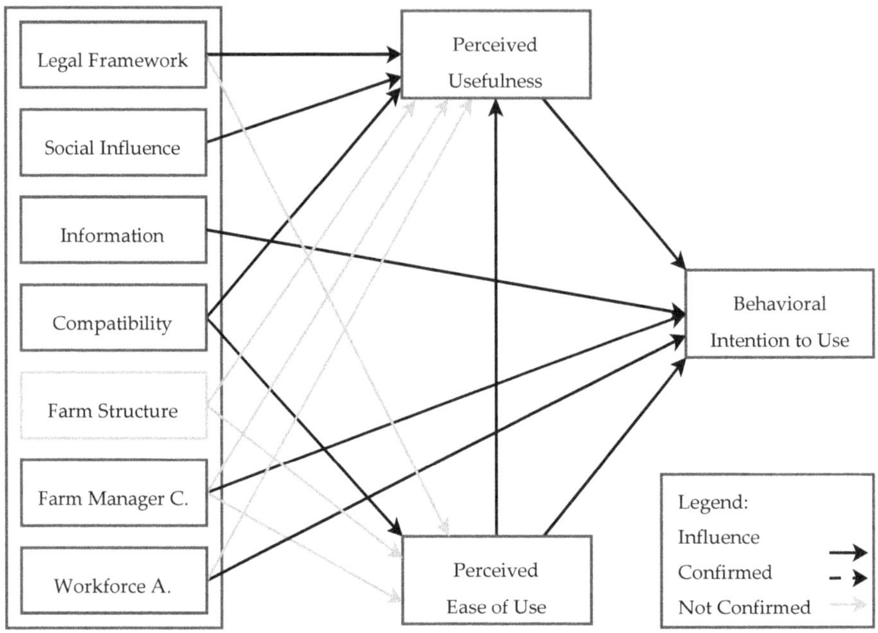

Figure 2. Examined TAM in the context of autonomous field robots.

3.2. Perceived Ease of Use

The perceived ease of use was confirmed by most experts as an important acceptance factor (E2, E3, E4, E7, E8, E9), and, for expert 8, it is actually the most important factor for farmers: *"But I think for farmers [the most important factor] is the complexity of the machines. (…) it's too difficult to use them."* (E8). For this reason, four of the respondents argue that AFR should be designed as simple and clean as possible (E3, E7, E8, E9). Expert 5, on the other hand, sees things completely differently: *"If you can use a smartphone, you can use a robot."* (E5). Experts 2 and 8 add that it is important to provide farmers with a training program, with expert 7 focusing mainly on practical demonstrations to help farmers overcome any fear of dealing with AFR. Even after the purchase of AFR, according to two experts, good support must be guaranteed so that farmers do not have to worry and to: *"(…) be afraid of not having the right support at the right time."* (E8). Therefore, it would be: *"(…) easier for the farmer accept, to take risk."* (E2).

For the TAM factor of perceived ease of use, the assumed influence on the intention of use was reflected in the answers of the respondents, too. The easier AFR are to operate, the more farmers are willing to use them, according to the experts. Hence, the indirect influence of perceived ease of use on perceived usefulness could be confirmed by the proposed training programs; the more farmers are trained about the possible applications of AFR, the more likely they are to recognize a possible benefit or added value of this technology (see Figure 2).

3.3. Legal Framework

All experts interviewed agreed on the fact that the legal framework strongly influences the acceptance and speed of adoption of AFR in the agricultural sector. In this context, experts 6, 7, and 10 call for general legislation on the handling of robots before regulating AFR in particular, especially with regard to safety issues: *"I think it definitely will cause a problem and we need new legislations for robots in general and especially for agriculture. Especially if it's a person working alongside the robot."* (E6) and:

"Which is also related to how we can technically ensure that we will not have an accident and this question is quite difficult to solve outdoors." (E7). So far, the legal framework is relative: *"(...) loosely defined what is allowed and what is illegal."* (E1), which is why mandatory standards would be helpful, otherwise every manufacturer would try to find a way around it: *"So, all the actors of robotics agriculture work together to find some compromise to put some robots on the field."* (E5). Due to the perceived lack of governmental support, experts 2, 4, and 8 assume that the diffusion rate of AFR in Europe will slow down, which will give other countries a head start: *"[it] is crossing different development rates in different countries. So, we see for example that Japan is releasing faster robots on the same level of development. Whereas Europe and the US are more restrictive, and they are more concerned about safety law. So probably we will see in Japan and Australia first prototypes hitting the market."* (E8). In addition to the legal requirements for AFR, three of the interviewees (E3, E7, E8) see a more restrictive legislation for the use of pesticides in agriculture, which could have a positive effect on the acceptance of AFR: *"On the other hand, however, the regulations, especially in the area of plant protection, are currently so drastically discussed that people say: 'I need to take the step now, I have to do this or crops will die'."* (E3).

The assumed effect of the legal framework on perceived usefulness was unanimously confirmed by the interviewed experts and at the same time, represents the basic requirement for a successful diffusion of this new technology. Not only the legal requirements for the use of AFR but also the tightening legal situation regarding the use of pesticides in agriculture seem to have an impact on acceptance by farmers. In addition, a new relationship of the legal framework to ease of use could be identified, since the sooner legal standards for AFR are introduced, the fewer compromises or legal grey areas have to be used by manufacturers and the easier AFR can be operated (see Figure 2).

3.4. Social Influence

Most respondents agree that social influence in the form of sociopolitical pressure has an impact on farmers' decisions (E2, E3, E4, E7, E8, E9), although expert 9 notes that: *"It used to be almost nothing, but it is changing really fast. I think the pressure from the government with regulation, but it mostly comes from the end-user. They want something more sustainable and I think this is pushing really hard on the farmers. So that is definitely something they are considering more and more and it's growing really, really fast."* (E9). The discussion about the use of pesticides in agriculture in particular sometimes seems to be driven more by emotion than by fact: *"We have conducted many, many interviews and we have had people [farmers] there who were almost close to tears. They say they've had enough."* (E3) or: *"But it's not always taking into account with very precise technical date, it's more passionate, you know."* and: *"(...) they [consumers] have this crusade against glyphosate."* (E8). According to one respondent (E7), this discussion has an influence on the acceptance of AFR among farmers, even before technical aspects, although in his opinion, the subject of labor safety should not be disregarded. Experts 2, 3, 4, 7, and 8 also see this sociopolitical discussion as a good opportunity for AFR to enter the market in order to meet the demands of society: *"If this solution [AFR] can give the farmer some positive recognition it will be more easy [to accept] for the farmer."* (E2).

The assumed relationship of social influence on perceived usefulness was clearly confirmed. AFR are seen by farmers as useful tools to counteract social pressure and the associated increasing regulation of pesticide use (see Figure 2).

3.5. Information

The vast majority of the respondents (E1, E2, E4, E5, E6, E7, E8, E10) agree that data protection and data autonomy play a role in AFR's acceptance process, of which most consider this role to be very important (E2, E4, E5, E7, E10). Thus, according to experts 2, 4, and 6, it is important for farmers that: *"(...) the data continue to belong to them [farmers] and that they have full control over the data."* (E4), because: *"(...) they want to be independent."* (E2). Several interviewees (E3, E5, E7, E8) see this rather in a twofold way: *"On the one hand, [some] say: 'yes, look, this is my field, my production, so to speak, and I would like to keep the pictures with me'. (...) On the other hand (...) they say: 'of course I'm very happy to share this information because I know that the system will then improve continuously, and I will benefit from sharing this information'."* (E3), which is also influenced by the type of farm, because in viticulture, for example, exists: *"(...) a culture of keeping secrets."* (E8) can be important for the success of the company (E7, E8). For one of the interviewees (E4), most farmers do not really understand the term data protection, which is why it is important for AFR's acceptance process to sell the issue to farmers in the right way. Expert 9, on the other hand, sees no influence of the information factor, because: *"(...) everyone is working with Google for example or e-mail for store your pictures, so everything is on Google, so it doesn't seem to matter for most of the people. Big companies yes, but small farmers no. Big companies will look at that, but I don't think there is anything to fear."* (E9).

It appears that the assumed influence of the factor information on behavioral intention was not just confirmed but also strengthened by the majority of the respondents. Only one expert (E9) saw it in a different light. However, the strength of this effect depends to a large extent on the farm manager characteristics and the farm structure, as the issue of data protection and data autonomy can be more important in viticulture, for example, than in arable farming, where the experts perceive a certain dichotomy among farmers (see Figure 2).

3.6. Compatibility

All the respondents agreed that AFR's compatibility with existing technology is very important for the perceived usefulness and ease of use among farmers. Because: *"(...) autonomous machines are the natural evolution of farm machinery."* (E8) and: *"(...) it would be nice if you could combine the past and the future."* (E3). Otherwise, it can be confusing for farmers sometimes (E3). According to Expert 6, AFR have to fit into the given farm structures and not vice versa: *"I think they [farmers] want to have a robot they can adapt to their own farm. They don't want to do it the other way around. They don't want to adapt their farm to the robot."* (E6), whereby this again depends on the farm size: *"like small farms that don't have necessarily like machinery and stuff you can adapt with their own habits. Not necessarily with the tractors. So, I think it depends on the farms, but if there is already a tractor it is easier for them to buy a robot, if the robot is adapted to the tractor, I think."* (E6). At the same time, three experts (E6, E8, E10) underlined that the existing AFR concepts are not yet compatible enough. Two of the respondents (E3, E8) also made it clear that in this process: *"(...) nevertheless the core know-how of these companies [established agricultural machinery manufacturers] must be taken along. It is unrealistic to expect that robotics manufacturers will suddenly revolutionize the seed planting process."* (E3).

In terms of the TAM factors, the assumed relationships of AFR's compatibility with perceived usefulness and perceived ease of use were confirmed. The better the new technology can be combined with the established technology, the less confusing it is for farmers and the easier it is to use. At the same time, the perceived usefulness of AFR is higher if it can be integrated easily into existing farm structures (see Figure 2).

3.7. Farm Manager Characteristics

There were differing views among the respondents on the role of the farmers' age in the acceptance process. Four experts (E1, E4, E6, E9) consider that: *"(...) age for example is not something that prevents people from buying a robot."* (E1). Expert 9 adds that: *"The elderly doesn't like technology most of the time, just like a smartphone they don't like it most of the time. But farmers are used to like technology in general, so I think the age doesn't matter."* (E9). Three of the interviewees (E2, E3, E7) share this view insofar as younger people do not necessarily show a higher level of acceptance than older people but are more sensitive or have a greater affinity for technology and that they: *"(...) consider that it's more like present. So, like this technology is for today and people a bit older or lower educated tend to think that it's maybe more the future."* (E7). On the other hand, experts 4 and 8 attribute an effect on acceptance of age by all means: *"Of course, younger farmers tend to be quicker in using such technology. I say, if you plot a curve, acceptance tends to increase with age."* (E4). One interviewee also sees an opportunity in AFR: *"(...) to have a new better image of agriculture and farming in general for a new and younger generation. I hope it would be that."* (E6), in order to counteract the increasing ageing in the sector.

With regard to the educational level of the farmers, the respondents were in agreement. For instance, about half of them (E2, E3, E4, E7, E9) were convinced that there is a correlation between level of education and AFR acceptance. Experts 3 and 8, however, tend to see AFR manufacturers as having a responsibility to design AFR with such simplicity that the level of education should not play a role. Two of the interviewees (E6, E8) find that there is no such influence at all. Instead, acceptance depends more on the farm structure, since: *"(...) tractors and combines, these machines are already sophisticated, so they [farmers] know how to use them."* (E8).

According to the experts, although the risk appetite or curiosity about innovative technologies plays a role in the intention to use AFR, this varies with the personality of the farmer (E3, E4, E6) and is sometimes correlated with the level of education (E1). One of the respondents therefore suggests that leasing concepts should be offered first, so that: *"(...) I do not have the entry barriers for the farmer. And then it's relatively easy, then the farmer actually has relatively little risk for the first year and that's the way it has to be."* (E3).

3.8. Workforce Availability

There is general agreement among the respondents that the workforce availability factor has an influence on the acceptance process. For example, all the experts surveyed see the increasing shortage of labor in agriculture as a driver for AFR's adoption. For two of the interviewees (E1, E7), the lack of skilled workers is even one of the most important drivers of acceptance: *"So, they have big issues finding people to drive the tractors and do the job. So, the main problem is this."* (E7). In addition, there is an increasing pressure from the trend to use less pesticides, which means more mechanical weed control measures, resulting in more labor-intensive weed control (E7). Three experts (E2, E5, E8) point out that for some physically demanding jobs in agriculture, you need to be in good shape or it is simply unhealthy, which is why you hardly find anyone to do it; therefore, there is no risk of a displacement effect: *"(...) nobody wants to do it, because it's very boring and very difficult (...) because we don't have someone. So, for me the robot won't have any impact on the labor market."* (E5). Experts 3, 6, 7, and 10, on the other hand, see an effect on the labor market in Eastern Europe, since the use of AFR means that fewer seasonal workers are needed, which could lead to a social problem: *"But if you imagine that like 80% of the labor is replaced by robots it's also a social question in terms of which society we want to build and how we will deal with that."* (E7) and: *"I think we're playing with new tools and not necessarily thinking about the impacts it's going to have. But we have to think about it of course."* (E6). Two of the respondents (E8, E9) counter that: *"(...) a new set of new jobs [is] coming with digital technologies, like data analyzer, like maintenance teams, so there will be jobs."* (E8). Finally, one expert points out that: *"(...) a person I hire to work in the field for an hour will always be more expensive than a robot."* (E4).

The assumed direct influence of workforce availability in the adapted TAM on the behavioral intention to use was clearly confirmed by the need to address labor shortages. In addition, a new relationship was found between workforce availability and perceived usefulness, because the less workers are available, the greater the perceived usefulness of AFR for farmers. This relationship is also confirmed by the fact that the removal of the driver reduces the cost of performing the task (see Figure 2).

4. Discussion and Conclusions

AFR are considered to be a promising technology to address, at least in part, the many problems in agriculture identified in this study. Nevertheless, there are hardly any studies to date that deal with the multidimensionality of AFR's ex ante user acceptance process. Such understanding is particularly important, in some cases existentially, for AgTech startups that have committed themselves to this new technology. Therefore, the aim of this explorative paper was to gain a first qualitative understanding of the factors influencing the ex ante user acceptance and diffusion of AFR from the perspective of AgTech startups to compare the results with the statements of farmers (as buyers) in a follow-up study. For this purpose, guideline-based interviews with ten European AgTech startups dealing with agricultural robotics were conducted at the FIRA in Toulouse (France). The interview guideline was based on the TAM according to [19] and was supplemented by external factors (legal framework, social influence, information, compatibility, farm manager characteristics, and workforce (availability)) taken from the scientific literature on this topic.

In conclusion, it was found that due to the scope of the topic, many factors influence the acceptance of farmers from the AgTech startups' point of view. In general, the assumed relationships in the TAM were not only confirmed, but also supplemented by a new factor (farm structure) and several new relationships, which once again demonstrates the transferability of the TAM to various problems of agricultural acceptance research (see also [37]).

Perceived usefulness was identified by the respondents as the most important or one of the most important factors influencing the ex ante user acceptance of AFR, with economic benefits ranking ahead of ecological ones. This finding is supported by the results of [10,38], who also investigated the ex ante user acceptance factors of automation and PA technologies based on a TAM. In this context, the price of AFR plays a role in so far as the return of investment must pay off in the end and not necessarily its absolute amount, which partly confirms the results of [11]. The author described high investment costs as a possible barrier to acceptance of AFR in a similarly qualitative survey. Therefore, the experts interviewed in this study see AFR first in crops with particularly high labor costs, such as special crops but also row crops, such as sugar beet, before they become economically attractive on a broader scale, which is also confirmed in other studies [2,8,39]. According to the experts, the ecological advantages of AFR, with the exception of organic farms, have a rather indirect effect on acceptance, as farmers are more or less forced to deal with such technologies due to the progressive reduction of legally permitted pesticide use, which is in line with [40]. At the same time, this sociopolitical pressure is an opportunity for AFR to enter the market more broadly. A similar conclusion was reached by the authors in [41] as a result of their systematic literature review. They particularly emphasized the importance of the ecological benefits of PA and the trend towards more sustainable agriculture. Reliability and safety issues of AFR were also considered by most respondents as an important part of perceived usefulness, which is in line with several other studies [3,5,42]. Overall, a better communication of the economic but also environmental benefits can contribute to an increase in ex ante user acceptance, which, regarding the reliability of AFR, should include practical demonstrations.

The respondents also attributed an important role in the diffusion process to ease of use. This confirms the studies of [24,25] on agriculture and on TAM in general [19]. In addition, Salimi et al. [10] have found that there is a significant positive correlation between automation features such as compatibility, low complexity, reduced workforce needs, etc., and perceived ease of use. According to the experts, AFR should be designed as simply as possible, which is in line with the findings of [3], according to which too much complexity of AFR can represent an important barrier to its diffusion.

The experts consider the legal framework to be a highly important factor and a major barrier to the diffusion of AFR in the EU. For example, the general operation of AFR is currently considered by both the experts and [40] to be legally uncertain and thus, declared a grey area. Dörr et al. [40] also point out that the regulations regarding the use of AFR vary widely around the globe, with Australia, Japan, and the USA being further along than Europe, where scientists tend to register a standstill. In the case of Germany, the federal government has at least stated its intention to establish the legal prerequisites for autonomous vehicles by the end of its term of office, to clarify liability issues [43]. A proposed solution to liability issues or to clarify the cause of an accident in connection with AFR could be the use of a "black box" analogous to aviation [44]. In addition, the experts emphasized a large, acceptance-influencing effect through the tightening of existing laws affecting agricultural practice; more precisely, by the ever more restrictive regulations on the use of pesticides in agriculture, which are forcing farmers to change. This effect is also confirmed by [40], who add that the future market share of autonomous machines will depend largely on regional legislation, while [2] add that legal framework may well have an influence on the economic benefits of AFR.

The main opinion of the interviewees that social influence in terms of social pressure for more sustainable farming practices also affects perceived usefulness and thus, indirectly the behavioral intention to use AFR has been confirmed several times in the scientific literature [45–47]. Zander et al. [48] conclude that farmers increasingly feel social criticism as a burden and are thus, influenced in their strategic decisions [29]. Devitt [5] argues that farmers may lose agricultural know-how in the medium as well as long term using AFR and fear a resulting loss of social reputation. This is confirmed by [10], who observed a significant negative correlation between social factors and perceived usefulness of automation adoption in Azerbaijan's agricultural sector. In contrast, a consumer survey in the EU showed that only six percent of the respondents would agree to a ban on robots in agriculture [49].

With respect to the influence of collected information by AFR, there is a large consensus among the experts that this is particularly important for the acceptance process among farmers, as on the one hand, their independence is important to them and on the other hand, they want to keep control of the machinery and certain business secrets (e.g., in viticulture). In their media analysis, Schleicher and Grandorfer [50] identified data protection and data autonomy as important barriers to acceptance of PA technologies in agriculture. In contrast, the experts surveyed in a study by [11] explicitly did not see data protection and data autonomy as barriers to the diffusion of AFR in agriculture. These contradictory results can possibly be explained by the dependence of the factor information on the characteristics of the farmer and the structure of the farm, as expressed by the experts interviewed in this survey.

The compatibility of AFR was also confirmed in its direct relevance to perceived usefulness and perceived ease of use and its indirect relevance to the behavioral intention to use. This is a particularly important transition phase from existing technologies to digital technologies in order to reach farmers accordingly and not to shy away from the feared complexity of the new system [11,51]. This is also confirmed by the results of [25], who conducted a quantitative survey among farmers on PA technology acceptance.

The farm structure, as a newly revealed influencing factor, seems to have a direct effect on perceived usefulness, as for example, AFR's ecological added value is higher for organic farms than for conventional farms. On the other hand, new technologies, such as AFR, can be better integrated into existing farm processes if the structural requirements (e.g., the degree of digitization) are met. Thus, Dörr et al. [40] see the farm structure factor as a possible reason for regional differences in AFR adoption. In addition, Shockley et al. [39] see an opportunity for smallholder farms in particular to gain economic benefits over large farms by using more scalable AFR, which is in contrast to the results for technology adoption in general, where larger farms tend to adopt new technology rather than small farms [52,53].

The farm manager's characteristics were seen as relevant to the acceptance process, despite some disagreement among experts. Thus, the findings of [5] that better trained and younger farmers are more willing to adopt AFR could only be partially confirmed. While the experts agreed with the influence of educational level, as confirmed by [12] for PA technologies or by [19] in general, they disagreed on the influence of age, as AFR should be designed in such a simple way that age is not relevant at all. Again, possible demonstrations of AFR could be helpful in overcoming possible barriers to acceptance due to the personal characteristics of the plant manager.

The availability of the workforce was also confirmed as an important factor influencing AFR's acceptance, the main argument being the existing shortage of skilled workers. This confirms the findings of [11], according to which the availability of workforce and rising wages will increase AFR's diffusion in the agricultural sector. The experts interviewed emphasized that there would be no major displacement of existing jobs, as there was already an existing shortage. The situation is different for seasonal workers from low-wage countries. In such countries, reference [30] therefore sees the introduction of AFR coming later than in countries with higher wage levels. Lowenberg-DeBoer et al. [2]

believe that migration flows and the resulting sociopolitical discussion also have an influence in this case. One expert agrees that replacing a driver with an AFR will always be cheaper than driver-operated machines. In addition, the experts see a contrary movement in the sociopolitical demand for a further reduction of pesticides, as this means more mechanical weed control, which in turn is more labor-intensive. However, respondents also noted that AFR will create new jobs, which is in line with [1], who argue that AFR could attract more highly qualified workers and thus, make agriculture more attractive to young people again.

Since these are empirically collected data, caution and certain limitations must be exercised in interpreting the results with regard to the examined ex ante user acceptance of AFR. First, the respondents are exclusively AgTech startups with AFR involvement, which may have an overly optimistic view of this new technology and their view on the acceptance of the farmers automatically shows not actual acceptance but their view on it. On the other hand, some of them also gained experience with diffusion barriers and the group of "early adopters" (e.g., Naïo Technologies as the market leader has already sold more than 120 AFR). Thus, a different sample composition could lead to different results [54]. Second, this is the ex ante user acceptance of a technology with which most farmers probably have had no contact so far and therefore, little can be said about the actual motivation for a possible adoption. Furthermore, the categorization of the expert statements to the different factors will always remain partly subjective. Causal relationships, as they can be calculated in e.g., quantitative studies using partial least squares (PLS) analysis, cannot be identified in this qualitative paper. Therefore, no hypotheses have been tested, as they cannot be statistically validated in terms of their significance.

However, the results offer various starting points for further research into the acceptance of autonomous machines in agriculture. Subsequently, the qualitative results must be quantified by surveying the farmers directly to verify or reject the results in a comparison, which is planned in a follow-up study. Furthermore, it would be interesting to investigate the conditions under which farmers would be willing to buy AFR (e.g., by using a choice experiment). Additionally, other stakeholder groups could be included in the analysis, for example, how consumers react to AFR depending on different designs. Furthermore, the results could support political decision-makers in dealing with this new technology (especially with regard to the creation of a contemporary legal framework for AFR) and AFR manufacturers in the promotion of their products among farmers. In conclusion, the findings of this and related studies can provide input for a successful implementation of AFR in agriculture in order to achieve the connected benefits for farmers, but most importantly for society in terms of a more sustainable and efficient farm management and thus, address the various social problems, as listed in the introduction.

Author Contributions: Conceptualization, F.R.v.V. and H.H.; methodology, F.R.v.V.; software, F.R.v.V.; validation, F.R.v.V. and H.H.; formal analysis, F.R.v.V.; investigation, F.R.v.V.; resources, F.R.v.V.; data curation, F.R.v.V.; writing—original draft preparation, F.R.v.V.; writing—review and editing, H.H.; visualization, F.R.v.V.; supervision, H.H.; project administration, H.H. All authors have read and agreed to the published version of the manuscript.

Funding: This research received no external funding.

Conflicts of Interest: The authors declare no conflict of interest.

Appendix A

Interview Guideline:

- Do you agree that the interview may be recorded and used in anonymous form for my research?
- What role will autonomous field robots play in agriculture over the next ten years?
- In your opinion, which factors inhibit/promote the acceptance of autonomous field robots among farmers?
- What influence do you attribute to the farm manager's age, educational level and risk propensity?

- Do you consider the aspect of data protection and data sovereignty to be important for the acceptance of autonomous field robots among farmers?
- In which areas of agriculture can you imagine the use of autonomous field robots?
- How important do you think it is that autonomous field robots are compatible with different tools and possibly conventional equipment?
- Do you expect the introduction of autonomous field robots to have an impact on the distribution structure in the agricultural machinery market?
- What effects do you see on the agricultural labor market if autonomous field robots are used as standard?
- How do you assess the ecological added value as an influencing factor on the acceptance among farmers?
- Can conflicts with current legislation be a problem for the introduction of autonomous field robots in agriculture?
- In summary, which factors do you think have the greatest influence on the adoption of autonomous robots in agriculture?
- When do you think the commercial use of autonomous field robots in agriculture will start?
- How much will the robots cost?

Table A1. Coding Guide.

Factor	Definition	Anchor Example	Coding Rule
Perceived Usefulness	- Economic efficiency (increase in yield or profit; input savings; investment costs) - Ecological sustainability - Reliability (reliable operation and performance during work peaks) - Reduced workload	*"(...) and that the farm is economically and ecologically in good shape."* (E3)	Statements on profitability, ecological added value, functionality, and performance expectations.
Perceived Ease of Use	- Handling/user interface - Own maintenance	*"The technology has to be simple anyway."* (E9)	Intensity of perceived ease of use's effect on acceptance.
Legal Framework	- Conflicts with law/liability (in the fields and on public roads) - legal restrictions on substitutes (e.g., for pesticides) - Subsidies	*"Yes, certainly. Because you have to adapt the law (...)"* (E10)	Legal framework as a requirement and driver for the adoption of AFR.
Social Influence	- Influence of social networks (e.g., society, consultants, colleagues)	*"They* [society] *want something more sustainable and I think this is pushing really hard on the farmers."* (E9)	Statements on the influence of social networks on the decision-making process of farmers.
Information	- Data processing and use of farmers actions for strategic production planning - Data protection and data autonomy	*"(...) everyone is working with Google (...) it doesn't seem to matter for most of the people. Big companies yes, but small farmers no."* (E9)	Intensity of the information effect on acceptance.

Table A1. *Cont.*

Factor	Definition	Anchor Example	Coding Rule
Compatibility	- Cooperation with existing technology	*"(...) a robot will be more important if it is connectable with all the other machines, (...)"* (E2)	Intensity of the compatibility effect on acceptance.
Farm Manager Characteristics	- Age, education, and risk appetite	*"(...) the farmers of today were not born with these technologies. (...) So yes it's preventing the use of digital technologies."* (E6)	Classification by age, educational level, and risk appetite. Intensity of the effect on acceptance.
Workforce Availability	- Influence of workforce issues (especially availability of workforce)	*"Regulations are one questions, but also safety on the work and the availability of workforce."* (E7)	Intensity of the workforce availability effect on acceptance.

References

1. Duckett, T.; Pearson, S.; Blackmore, S.; Grieve, B. Agricultural robotics: The future of robotic agriculture. *arXiv* **2018**, arXiv:1806.06762.
2. Lowenberg-DeBoer, J.; Huang, I.Y.; Grigoriadis, V.; Blackmore, S. Economics of robots and automation in field crop production. *Prec. Agric.* **2019**, *21*, 278–299. [CrossRef]
3. Redhead, F.; Snow, S.; Vyas, D.; Bawden, O.; Russell, R.; Perez, T.; Brereton, M. Bringing the farmer perspective to agricultural robots. In Proceedings of the CHI Conference on Human Factors in Computing Systems, Seoul, Korea, 18–23 April 2015; Association for Computing Machinery: New York, NY, USA, 2015. [CrossRef]
4. King, A. Technology: The future of agriculture. *Nature* **2017**, *544*, 21–23. [CrossRef] [PubMed]
5. Devitt, S.K. Cognitive factors that affect the adoption of autonomous agriculture. *Farm Policy J.* **2018**, *15*, 49–60. [CrossRef]
6. Kester, C.; Griepentrog, H.W.; Hörner, R.; Tuncer, Z. A survey of future farm automation—A descriptive analysis of survey responses. In *Precision Agriculture '13*; Stafford, J.V., Ed.; Wageningen Academic Publishers: Wageningen, The Netherlands, 2013; pp. 785–792. [CrossRef]
7. Erl, T.; Puttini, R.; Mahmood, Z. *Cloud Computing: Concepts, Technology & Architecture*, 1st ed.; Prentice Hall: Upper Saddle River, NJ, USA, 2013.
8. Minßen, T.-F.; Urso, L.-M.; Gaus, C.-C.; Frerichs, L. New plant production systems with autonomous agricultural machinery. *ATZoffhighway* **2015**, *8*, 6–11. [CrossRef]
9. Burwood-Taylor, L.; Leclerc, R.; Tilney, M. AgTech Investing Report: Year in Review 2015. *AgFunder*. Available online: https://research.agfunder.com/2015/AgFunder-AgTech-Investing-Report-2015.pdf (accessed on 31 October 2020).
10. Salimi, M.; Pourdarbani, R.; Nouri, B.A. Factors affecting the adoption of agricultural automation using Davis's acceptance model (case study: Ardabil). *Acta Technol. Agric.* **2020**, *23*, 30–39. [CrossRef]
11. Rial-Lovera, K. Agricultural robots: Drivers, barriers, and opportunities for adoption. In Proceedings of the 14th International Conference on Precision Agriculture, Montreal, QC, Canada, 24–27 June 2018; The International Society of Precision Agriculture: Monticello, IL, USA, 2018.
12. Caffaro, F.; Cavallo, E. The effects of individual variables, farming system characteristics and perceived barriers on actual use of smart farming technologies: Evidence from the Piedmont region, northwestern Italy. *Agriculture* **2019**, *9*, 111. [CrossRef]
13. Bukchin, S.; Kerret, D. Food for hope: The role of personal resources in farmers' adoption of green technology. *Sustainability* **2018**, *10*, 1615. [CrossRef]
14. Pathak, H.S.; Brown, P.; Best, T. A systematic literature review of the factors affecting the precision agriculture adoption process. *Prec. Agric.* **2019**, *13*, 1–25. [CrossRef]
15. Rogers, E.M. *Diffusion of Innovations*, 5th ed.; Free Press: New York, NY, USA, 2003.
16. Mayring, P. *Qualitative Inhaltsanalyse: Grundlagen und Techniken*, 12th ed.; Beltz: Weinheim, Baden-Württemberg, Germany, 2015.
17. Die schriftliche Befragung. Available online: https://www.hb.fh-muenster.de/opus4/frontdoor/deliver/index/docId/489/file/SchriftlicheBefragung.pdf (accessed on 28 November 2020).

18. Gläser, J.; Laudel, G. *Experteninterviews und Qualitative Inhaltsanalyse als Instrumente*, 4th ed.; VS Verlag für Sozialwissenschaften: Hesse, Germany, 2010.
19. Davis, F.D. Perceived usefulness, perceived ease of use, and user acceptance of information technology. *MIS Q.* **1989**, *13*, 319–340. [CrossRef]
20. Chuttur, M. Overview of the technology acceptance model. Origins, developments and future directions. *Sprouts Work. Papers Inf. Syst.* **2009**, *9*, 1–21.
21. Miller, L.H. A Study in man-machine interaction. In Proceedings of the National Computer Conference, Dallas, TX, USA, 13–16 June 1977; Association for Computer Machinery: New York, NY, USA, 1977. [CrossRef]
22. Benbasat, I.; Dexter, A.S. An investigation of the effectiveness of color and graphical information presentation under varying time constraints. *MIS Q.* **1986**, *10*, 59–83. [CrossRef]
23. Verma, P.; Sinha, N. Integrating perceived economic wellbeing to technology acceptance model: The case of mobile based agricultural extension service. *Technol. Forecast. Soc. Change* **2018**, *126*, 207–216. [CrossRef]
24. Adrian, A.M.; Norwood, S.H.; Mask, P.L. Producers' perceptions and attitudes toward precision agriculture technologies. *Comput. Electron. Agric.* **2005**, *48*, 256–271. [CrossRef]
25. Aubert, B.A.; Schroeder, A.; Grimaudo, J. IT as enabler of sustainable farming: An empirical analysis of farmers' adoption decision of precision agriculture technology. *Dec. Support Syst.* **2012**, *54*, 510–520. [CrossRef]
26. Paustian, M.; Theuvsen, L. Adoption of precision agriculture technologies by German crop farmers. *Prec. Agric.* **2016**, *18*, 701–716. [CrossRef]
27. 2035 Arable Farming Strategy—Prospects for Productive and Diverse Crop Farming. Available online: https://www.bmel.de/SharedDocs/Downloads/EN/Publications/ackerbaustrategie-en.pdf?__blob=publicationFile&v=6 (accessed on 6 June 2020).
28. Alt, N. Gesetzliche Vorgaben für den Feldeinsatz von hochautomatisierten Landmaschinen. In *Jahrbuch Agrartechnik 2018*, 1st ed.; Frerichs, L., Ed.; Institute for Mobile Machines and Commercial Vehicles: Brunswick, Germany, 2018; pp. 1–7.
29. Kuczera, C. *Der Einfluss des Sozialen Umfeldes auf Betriebliche Entscheidungen von Landwirten*, 1st ed.; Margraf Publishers: Weikersheim, Germany, 2006; pp. 56–171.
30. Griepentrog, H.W. Automatisierung in der Außenwirtschaft. In *Automatisierung und Roboter in der Landwirtschaft*, 1st ed.; Molnar, C., Pikart-Müller, M., Eds.; Kuratorium für Technik und Bauwesen in der Landwirtschaft: Darmstadt, Germany, 2010; pp. 25–33.
31. Metschke, R.; Wellbrock, R. *Datenschutz in Wissenschaft und Forschung*, 3rd ed.; Druckerei Conrad GmbH: Berlin, Germany, 2002.
32. Weeding Service for Organic Farms. Available online: https://www.farming-revolution.com/ (accessed on 30 October 2020).
33. Ecorobotix Successfully Raises CHF 10.6 Million in Series B Financing with Lead Investors Capagro and BASF Venture Capital (in Englisch). Available online: https://www.basf.com/global/de/who-we-are/organization/locations/europe/german-companies/BASF_Venture-Capital/publications/2018/P-18-530.html (accessed on 30 October 2020).
34. La Robotique Au Service de l'Agriculture. Available online: https://www.naio-technologies.com/ (accessed on 1 July 2020).
35. RObotics for MIcrofarms. Available online: https://cordis.europa.eu/project/id/773875/de (accessed on 1 July 2020).
36. Intelligent Decision from Vineyard Robots. Available online: https://cordis.europa.eu/project/id/737669/de (accessed on 1 July 2020).
37. Alambaigi, A.; Ahangari, I. Technology acceptance model (TAM) as a predictor model for explaining agricultural experts behavior in acceptance of ICT. *Int. J. Agric. Manag. Develop.* **2015**, *6*, 235–247. [CrossRef]
38. Pierpaoli, E.; Carli, G.; Pignatti, E.; Canavari, M. Drivers of precision agriculture technologies adoption: A literature review. *Proc. Technol.* **2013**, *8*, 61–69. [CrossRef]
39. Shockley, J.M.; Dillon, C.R.; Shearer, S.A. An economic feasibility assessment of autonomous field machinery in grain crop production. *Prec. Agric.* **2019**, *20*, 1068–1085. [CrossRef]

40. Dörr, J.; Fairclough, B.; Henningsen, J.; Jahić, J.; Kersting, S.; Menning, P.; Peper, C.; Scholten-Buschhoff, F. *Scouting the Autonomous Agricultural Machinery Market*; Fraunhofer IESE: Kaiserslautern, Germany; Kleffmann Group: Lüdinghausen, Germany, 2019. Available online: https://www.iese.fraunhofer.de/content/dam/iese/en/dokumente/smart-farming/Scouting-the-Autonomous-Agricultural-Machinery-Market.pdf (accessed on 31 October 2020).
41. Finger, R.; Swinton, S.; El Benni, N.; Walter, E. Precision farming at the nexus of agricultural production and the environment. *Ann. Rev. Res. Econ.* **2019**, *11*, 313–335. [CrossRef]
42. Bechar, A.; Vigneault, C. Agricultural robots for field operations: Concepts and components. *Biosyst. Eng.* **2016**, *149*, 94–111. [CrossRef]
43. Koalitionsvertrag zwischen CDU, CSU und SPD. 19. Legislaturperiode. Available online: https://www.bundesregierung.de/resource/blob/656734/847984/5b8bc23590d4cb2892b31c987ad672b7/2018-03-14-koalitionsvertrag-data.pdf?download=1 (accessed on 1 July 2020).
44. Recht für funktionale Sicherheit in der Autonomik. Available online: https://www.digitale-technologien.de/DT/Redaktion/DE/Downloads/Publikation/autonomik-Leitfaden2.pdf?__blob=publicationFile&v=3 (accessed on 29 October 2020).
45. Foster, A.D.; Rosenzweig, M.R. Learning by doing and learning from others: Human capital and technical change in agriculture. *J. Politic. Econ.* **1995**, *103*, 1176–1209. [CrossRef]
46. Venkatesh, V.; Davis, F.D. A Theoretical extension of the technology acceptance model: Four longitudinal field studies. *Manage. Sci.* **2000**, *45*, 186–204. [CrossRef]
47. Venkatesh, V.; Morris, M.G.; Davis, G.B.; Davis, F.D. User acceptance of information technology: Toward a unified view. *MIS Q.* **2003**, *27*, 425–478. [CrossRef]
48. Zander, K.; Isermeyer, F.; Brügelt, D.; Christoph-Schulz, I.; Salamon, P.; Weible, D. *Erwartungen der Gesellschaft an die Landwirtschaft*; Stiftung Westfälische Landschaft; Johann Heinrich von Thünen-Institut: Brunswick, Germany, 2013.
49. Public Attitudes Towards Robots. Special Eurobarometer 382. Available online: https://ec.europa.eu/commfrontoffice/publicopinion/index.cfm/Survey/getSurveyDetail/instrumen687ts/SPECIAL/surveyKy/1044/p/3 (accessed on 1 July 2020).
50. Schleicher, S.; Grandorfer, M. Digitalisierung in der Landwirtschaft: Eine Analyse der Akzeptanzhemmnisse. In *Digitale Marktplätze und Plattformen, Lecture Notes in Informatics 278*; Ruckelshausen, A., Meyer-Aurich, K., Borchardt, C., Hofacker, J.-P., Loys, R., Schwerdtfeger, H.-H., Sundermeier, H., Theuvsen, F.B., Eds.; Gesellschaft für Informatik: Bonn, Germany, 2018; pp. 203–206.
51. Trendov, N.M.; Varas, S.; Zenf, M. *Digital Technologies in Agriculture and Rural Areas: Status Report*; Food and Agricultural Organization of the United Nations: Rome, Italy, 2019; pp. 118–120.
52. Anderson, P.D.; Wilson, N.P.; Thompson, D.G. Adoption and diffusion of level fields and basins. *J. Agric. Res. Econ.* **1999**, *24*, 186–203.
53. Bjornlund, H.; Nicol, L.; Klein, K.K. The adoption of improved irrigation technology and management practices—A study of two irrigation districts in Alberta, Canada. *Agric. Water Manag.* **2009**, *96*, 121–131. [CrossRef]
54. Melenhorst, A.-S.; Bouwhuis, D.G. When do older adults consider the internet? An exploratory study of benefit perception. *Gerontechnology* **2004**, *3*, 89–101. [CrossRef]

sustainability

MDPI

Article

Determination of Dehumidification Capacity of Water Wall with Controlled Water Temperature: Experimental Verification under Laboratory Conditions

Katarina Cakyova [1,2,*], Frantisek Vranay [3], Marian Vertal [3] and Zuzana Vranayova [3]

1 Center for Research and Innovation in Construction, Faculty of Civil Engineering, Technical University of Košice, 042 00 Košice, Slovakia
2 Institute of Building Structures, Faculty of Civil Engineering, Brno University of Technology, 602 00 Brno, Czech Republic
3 Institute of Architectural Engineering, Faculty of Civil Engineering, Technical University of Košice, 042 00 Košice, Slovakia; frantisek.vranay@tuke.sk (F.V.); marian.vertal@tuke.sk (M.V.); zuzana.vranayova@tuke.sk (Z.V.)
* Correspondence: katarina.cakyova@tuke.sk; Tel.: +421-944-731-360

Abstract: Water elements with flowing water on the surface are common in buildings as a form of indoor decoration, and they are most often perceived as passive humidifiers. However, by controlling water temperature, they can be also used for air dehumidification. The dehumidification capacity of indoor water elements was investigated experimentally under laboratory conditions. For the experimental verification of dehumidification capacity, a water wall prototype with an effective area of falling water film of 1 m^2 and a measuring system were designed and developed. A total of 15 measurements were carried out with air temperatures ranging from 22.1 °C to 32.5 °C and relative humidity from 58.9% to 85.6%. The observed dehumidification capacity varied in the range of 21.99–315.36 g/h for the tested measurements. The results show that the condensation rate is a dynamic process, and the dehumidification capacity of a water wall strongly depends on indoor air parameters (air humidity and temperature). To determine the dehumidification capacity of a water wall for any boundary conditions, the equations were determined based on measured data, and two methods were used: the linear dependence between humidity ratio and condensation rate, and nonlinear surface fitting based on the dependence between the condensation rate, air temperature, and relative humidity.

Keywords: water wall; falling water film; relative humidity; temperature; condensation rate; dehumidification

Citation: Cakyova, K.; Vranay, F.; Vertal, M.; Vranayova, Z. Determination of Dehumidification Capacity of Water Wall with Controlled Water Temperature: Experimental Verification under Laboratory Conditions. *Sustainability* **2021**, *13*, 5684. https://doi.org/10.3390/su13105684

Academic Editor: Mostafa Ghasemi Baboli

Received: 19 April 2021
Accepted: 15 May 2021
Published: 19 May 2021

1. Introduction

In the field of the research and development of water walls for building applications, water walls where water flows between two solid materials are mainly presented. This is a closed system and is used to adjust thermal comfort, especially air temperature. The first water wall of this type reported in the available literature was the one built in 1947 at the Massachusetts Institute of Technology by Hoyt Hottel and his students with the aim to use a water wall as a passive solar system [1]. It has been proven that a water wall system used as a green building façade can enhance the energy performance [2–5] and also fire protection of buildings [5].

Another group of water walls (Figure 1) represents decorative water features often located in the interior of public spaces such as shopping centers, hospitals, libraries, the entrance areas of office buildings, and hotels, but also airports. In this case, water falls on solid material and is in direct contact with indoor air, so there is an assumption that in addition to thermal comfort, other parameters of the indoor environment can also be affected. Water elements similar to green walls are perceived as a significant aesthetic element of the indoor environment and represent one of the patterns of biophilic design [6–8].

(**a**) (**b**) (**c**) (**d**)

Figure 1. Water wall examples: (**a**) Shopping center, Vienna, Austria; (**b**) Library, Brno, Czech Republic; (**c**) Railway station, Vienna, Austria; (**d**) Office building, Baku, Azerbaijan.

In the case of indoor greenery (e.g., living walls), many studies have been conducted that define the different systems and designs of these elements [9–12]. It was shown that indoor greenery offers several benefits, such as producing oxygen through photosynthesis [13]; affecting the physical parameters of indoor air quality, mainly temperature and humidity [14–16], CO_2 concentration, and acoustics [13,15]; and providing an aesthetically pleasing environment that increases productivity and lowers stress in the building's occupants [17]. The use of indoor plants in some situations could also be a potential problem, however, mainly in the context of humidity and fungal respiratory diseases in patients [10,14].

On the other hand, the issue of indoor water bodies, which also represent the possibility of bringing nature inside a building, has not been explored in detail. There is no clear division that describes these elements, even though water elements are increasingly becoming part of the interior of buildings. Water elements in the interior of a building enhance the experience of a place through the seeing, hearing, or touching of water and are associated with reducing stress, increasing feelings of tranquility, and lowering heart rate and blood pressure, as well as improving concentration and memory restoration [8,18]. It was found that the sound of water has been positively associated with restoration [19], especially in connection with dissatisfaction with environmental noise through windows [20]. However, there are also differences in the way different genders respond to water features, with male subjects tending to respond to water features more frequently than female subjects [18].

At present, water walls where the water circulates are mostly used for an evaporation cooling effect, as a consequence of which the air temperature decreases and water temperature increases [1,21,22]. In addition to temperature, air humidity also changes and increases. This effect may cause an unpleasant level of thermal comfort, especially in hot and humid conditions. Subsequent air treatment and dehumidification may also result in significant energy loads. A study from India showed that direct evaporation cooling was not as effective during the months of July and August, and some dehumidification is required to maintain thermal comfort indoors [23]. Research from Malaysia (which has a hot and humid climate) showed hybrid cooling strategies—thermal stack flue, cross ventilation, and water walls (evaporative cooling)—located in the atriums of office buildings. Due to use of this hybrid system, indoor temperatures were lower than outdoor temperatures, and indoor humidity fluctuated from 68% to 92% during working hours. This system does not increase RH significantly, but this value can still cause some level of discomfort [24]. The high humidity of air can also lead to problems associated with condensation, mainly in the case of air conditioners without regulation of humidity—for example, radiant cooling systems [25,26]. The evaporative cooling potential of water walls is more suitable for hot and dry climates. However, there is still a risk of Legionella bacteria with increasing water temperature [27]. Additionally, there is an assumption that regulation of water temperature can lead to a useful impact on indoor air temperature and humidity [28]. Research conducted by Fang et al. [29] showed the potential of using water walls with water temperature regulation. However, in this case, the experimental measurements took place in real conditions without control of the physical parameters of the inner microclimate in a specific environment of sorption-active elements. Thus, the resulting effect of the water element on the environment is closely related to the boundary conditions for the given environment; additionally, only the change in individual physical parameters was described, while the share of the water wall for this change was not clearly quantified.

At present, many studies deal with energy consumption in buildings, especially in the context of reaching the required temperature [30–32]. However, achieving the required humidity level is also important. Today, around 10–15% of the total energy consumed by a building is used to achieve the required humidity level [33]. Moreover, based on climate change and the global warming effect, there is an assumption that this percentage will increase several times in the coming years [34,35]. Similarly, as water changes the hydrothermal behavior of buildings with green components [36], the presence of water elements in interior spaces can affect the parameters of the indoor microclimate. This forces us to effectively utilize all elements and to learn more about the potential of individual elements in the building.

The presented paper fills the gap of knowledge in the field of effective use of water walls with controlled water temperature, with a focus on the dehumidification capacity of water walls that is defined by condensation rate under varying boundary conditions. The novelty of the contribution lies mainly in the fact that the water wall and its dehumidification potential are determined in laboratory conditions at various boundary conditions without sorption-active materials. The main object is the quantification of water wall dehumidification capacity via equations of condensation rate that can be used for different spaces of buildings. The paper connects the aesthetic element (water wall) with the functional use (dehumidification) in the interior of buildings. Quantification of dehumidification capacity of water walls with controlled water temperature as a device through an equation is helpful for its design, operation, and implementation in any space.

2. Materials and Methods

2.1. Research Scheme

Experimental verification under laboratory conditions was chosen to quantify the dehumidification potential of water walls. In Figure 2, it is possible to see a schematic illustration of the methodology of the work. Firstly, the assumption of water wall use was defined. The second step was to design and create a full-scale prototype of a water wall, and then the laboratory conditions were determined and specified. The next step was to perform measurements under different boundary conditions and analysis of the obtained data where condensation rate is expressed by incensement of weight condensate in time. Finally, the co-dependence of parameters (air temperature and humidity) and condensation rate was evaluated using two methods—linear and nonlinear analysis. The result is a comparison of both methods and the definition of water walls with controlled water temperature as a device with an impact on the humidity of the indoor environment. The Microsoft Excel 2016 and OriginPro 9.0 programs were used for the analysis and presentation of the results.

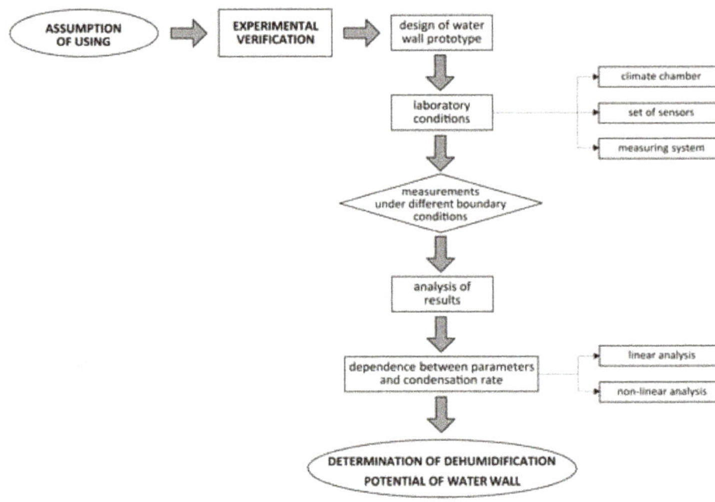

Figure 2. Research scheme.

2.2. Assumption of Water Wall Use to Dehumidify Air

For the effective use of water walls with a controlled water temperature, the phase change between liquid and gas associated with condensation was used. If water vapor comes into contact with a surface whose temperature is below the dew point of the air, the condensation process occurs, and the vapor changes phase and becomes liquid. In the present paper, the surface with a temperature below the dew point that was used is the falling water film with an average temperature of 14 °C. This process decreases the content of water vapor in the air.

In addition to mass transfer, heat transfer also occurs. Considering the problem of conjugated heat and mass transfer, three basic elements of the system can be specified: the solid body (glass), the thin falling water film flowing down the glass, and the surroundings of the gas mixture (air) [37]. The study of film flow is difficult because the flow itself exhibits large uncertainties and instabilities. Such large uncertainties can be attributed to its diverse, complex, and irregular wave structures [38]. At the top of a falling film unit, i.e., at the liquid inlet, the film is smooth. As the film flows downwards, there is a simultaneous development of hydrodynamic profiles [39]. During the flow of the water film, a characteristic wave pattern is created, which breaks into wave segments with different sizes and shapes. The wave pattern also depends on the flow velocity of the surrounding air [40]. Many studies [37,39,41,42] have proven that film thickness is one of the most essential parameters playing a vital role in determining heat and mass transfer performance. The mass flow rate, but also the shape, surface roughness, or porosity of the solid material after which it flows, have a significant influence on the formation and thickness of the water film and its energy benefit [43,44]. The problem of the falling water film is most often studied theoretically via numerical analysis [37] or via an experimental investigation [42].

2.3. Design of Water Wall Prototype

There are various designs and construction methods of decorative water features. In this paper, the construction method based on an overflow edge was used. For the experimental verification, the full-scale water wall prototype was designed (Figure 3). The prototype consists of bottom and upper water tanks with rectangular shapes made of polypropylene material (thickness 6 mm) using a method based on fusion welding; this ensured the tanks were watertight. From the bottom tank, the water is pumped and supplied to the upper water tank, and subsequently, the water spills over the overflow edge, then flows down the glass and forms a continuous falling water film on it. The effective area of the water film is 1 m^2; the dimensions are 710 mm x 1420 mm, and thus, the width to height ratio is 1:2.

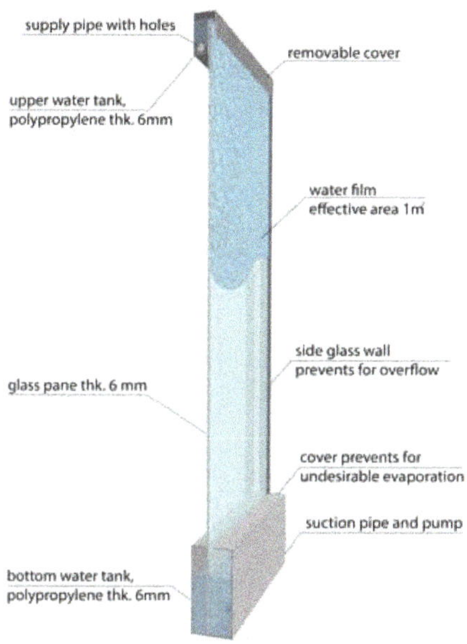

Figure 3. Design of water wall prototype.

The formation and thickness of the water film affect, among other things, the geometry of the upper part of the water wall (Figure 4). Into the collecting tank (1) is inserted a perforated pipe (2) for the supply of water (8). The length of the perforated pipe (2) is 700 mm, and the diameter of the holes is 3 mm, with an axial distance of 10 mm (Figure 5a). The orientation of perforation is downwards, which is essential for a smooth inlet of water to the tank and for forming a water film along the entire width of the water wall. The glass pane (4) forms the overflow edge (9) and is supplemented by protruding parts at its side edges to prevent undesired water overflow. A silicone adhesive (7) was used to connect the collecting tank (1) and glass pane (4). A ball valve (10) fastened to the collecting tank (1) serves to empty it if necessary. The collecting tank (1) is anchored to the supporting metal structure (5) by screws (6) and is supplemented by a cover (3). A cover is used also for the bottom tank to prevent unwanted evaporation. The solution of the upper part of the water wall is part of utility model PUV 50129-2017 no. 8710 [45].

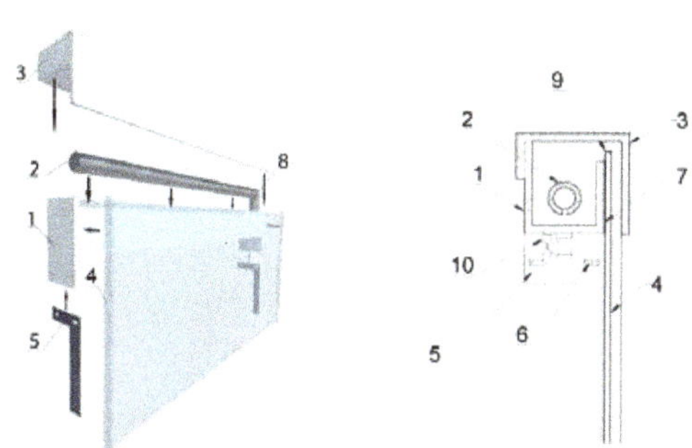

Figure 4. Design of water wall prototype.

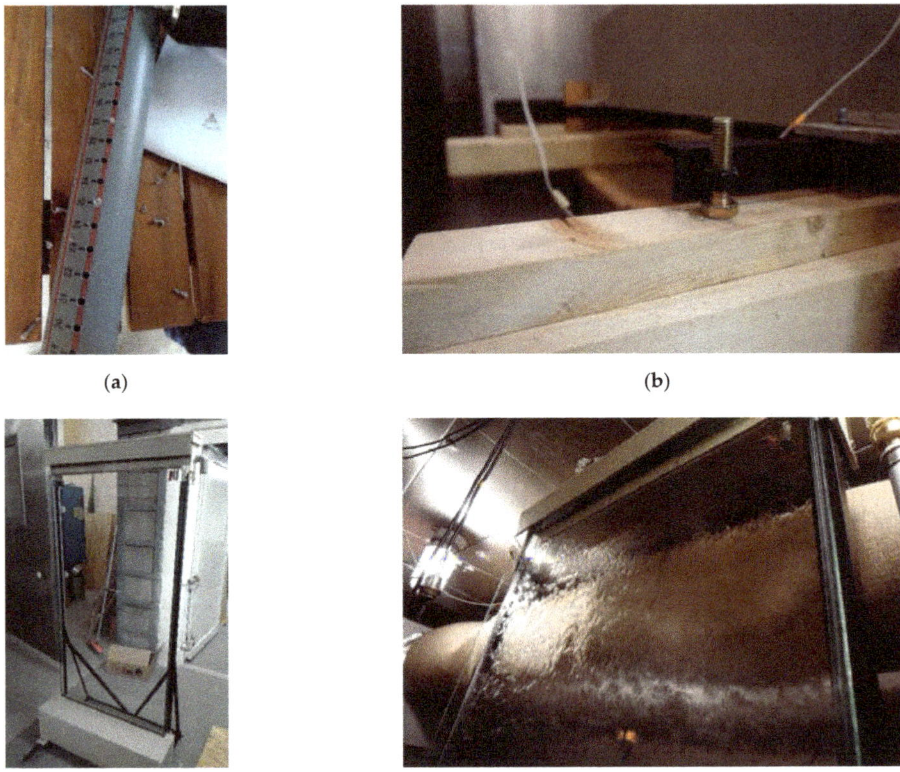

(a) (b) (c) (d)

Figure 5. Parts of water wall prototype: (**a**) Perforated pipe; (**b**) Screws for the horizontal position; (**c**) Final water wall prototype; (**d**) Water film created on the glass pane.

Furthermore, to form a water film along the entire width of the water wall, the overflow edge must be in a perfectly horizontal position. For this purpose, there are 4 screws located in the lower part of the metal structure (Figure 5).

2.4. Laboratory Conditions

2.4.1. Climate Chamber

The water wall prototype was tested under laboratory conditions using a climate chamber (Figure 6) located in the laboratory of the Faculty of Civil Engineering of the Technical University of Košice, where it is possible to set steady boundary conditions (air temperature and humidity can be controlled). The chamber is fully closed, airtight, and consists of two separate rooms. The air temperature and humidity can be controlled for each room separately. For the experiment, just one room was used, with inner dimensions of 3.95 × 1.60 × 2.85 m and volume of air of 18.01 m^3. The temperature ranged from −20 °C to +125 °C, and the RH of the chamber was adjustable from 20% to 95%. The climatic chamber was chosen to avoid mass exchange between the chamber's walls, which are made of stainless steel, and air. Due to this, the moisture buffer effect of materials can be ignored, and the clean dehumidification capacity of the water wall can be determined. The climate chamber was adjusted and adapted for measurement purposes. The modification of the chamber involved the creation of a textile tunnel that was deployed to the air supply. This intervention contributed to a uniform air distribution in the chamber and, at the same time, avoided a heavy draft.

2.4.2. Set of Sensors

In the chamber was installed a set of measuring sensors, which recorded boundary conditions throughout all the experiments. To measure the water temperature, the Ahlborn NTC sensor FN 0001 K was used. Two sensors measured the water temperature in the bottom tank and two in the upper tank. Moreover, two sensors type Ahlborn FHAD 46-C0 were installed to measure the temperature and the relative humidity of the air in front of the water film, and one behind the glass. The limiting deviations of the used sensors are described in Table 1.

(a) **(b)**

Figure 6. Climate chamber: (**a**) Exterior view; (**b**) Interior view.

Table 1. Specifications of measuring sensors.

Parameter		Device	Range	Deviation
Water	temperature	NTC sensor FN 0001 K	0 to 70 °C	±0.2 K
Air	humidity	FHAD 46-C0	10 to 90% RH	±2.0% RH at nominal temperature (+23 °C)
	temperature		5 to 60 °C	typical ±0.2 K (maximum ±0.4 K)

All the sensors were calibrated before use and were connected to a control unit (type Ahlborn) for the storage of the measured parameters. The AMR Win Control software was used to set up experiments and collect the measured data. This ensured the continuous recording of the measured values.

2.4.3. Measuring Systems

The water wall prototype was adapted for experimental testing in accordance with the scheme (Figure 7). The prototype was placed on an elevated wooden construction, the bottom water tank was connected with a measuring cylinder, and the different types of components, i.e., valves, flow meter, and plate heat exchanger, were added to the system.

Because there is no standard method for measuring the condensation rate [46], a system based on the weighting of condensate increment measurement was developed (Figure 8). For this purpose, the Radwag laboratory scale connected to a measuring cylinder was used. Its weight limit is 6000 g to an accuracy of 0.01 g. Another important restriction is the fact that ambient relative humidity should not exceed 85%. Same as the sensors, the laboratory scale was connected to the Ahlborn control unit. To achieve the right conditions for condensation, the plate heat exchanger was added to the system and connected to well water. Due to this, the water temperature was constant with a value of around 14 °C. After the start of the experiment, the individual parts of the water wall are flooded, and water needs to be added to the level of overflow created in the measuring cylinder. Subsequently, water from the bottom water tank flows into the measuring cylinder. From it, water is supplied into the upper part of the water wall by the pump and cooled by the plate exchanger. During the experiment, the water increments caused by the condensation process on the falling water film overflow to the plastic collecting tank, which was put on a laboratory scale. The flow rate velocity is controlled by the shut-off valve, and for accurate monitoring, the flow meter was used. Throughout all the experiments, the flow rate was 450 l/h, which was determined to be the optimal flow rate during past experiments [47]. Via this system, it is possible to accurately and continuously record the condensation rate (condensate formed on the water film) over time.

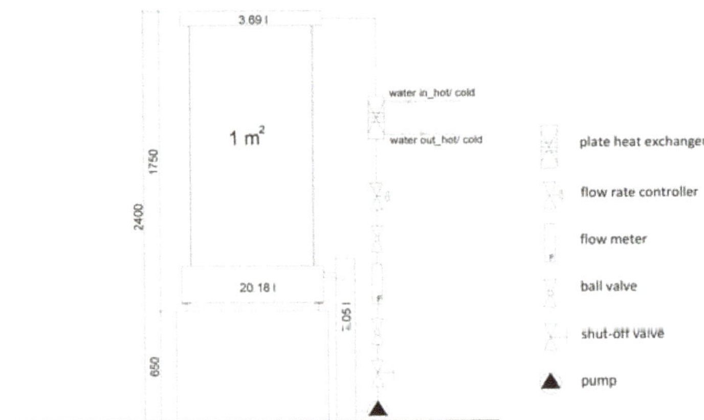

Figure 7. Schematic diagram of basic setup of the measurement system.

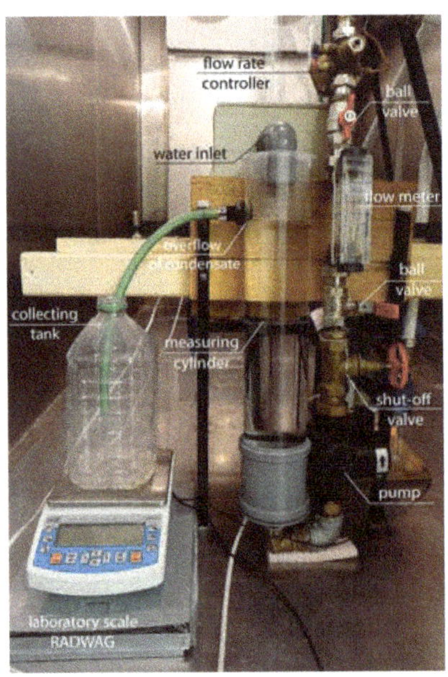

(**a**)

(**b**)

Figure 8. The measuring system for dehumidification performance: (**a**) Detailed view of the system; (**b**) Overall view of the system and the water wall located in the climate chamber.

3. Results and Discussion

A total of 15 measurements were carried out, with 2 repetitions to determine the dehumidification capacity of the water wall. During the measurements, air temperature varied from 22.1 °C to 32.5 °C, and relative humidity changed from 58.9% to 85.6% (Figure 9).

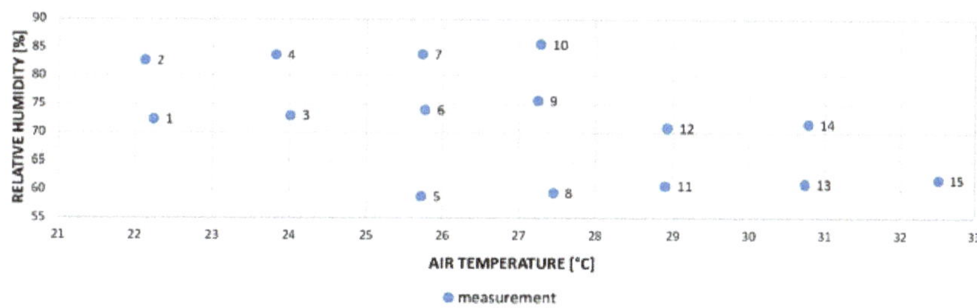

Figure 9. Scatter plot of boundary conditions during measurements.

The recording of measurements was performed in a 5 min time step. To avoid exceeding the measuring limit of the laboratory scale, the duration of the experiments ranged from 7 to 15 h. The obtained data from sensors were analyzed and evaluated to average values for each measurement. This means that an average water temperature, average air temperature, and average relative humidity were determined. The average water temperature was 14.8 °C ± 0.3 in the bottom tank and 14.0 °C ± 0.2 in the upper tank. To determine the condensation rate, for each measurement, the weight of condensate in time was evaluated. The resulting values of condensation performance of the water wall prototype and boundary conditions are evaluated in Table 2. This table also shows that the condition for condensation has been met: the water temperature must be lower than the dew point temperature of the air. The dew point temperature (T_{dp}) was calculated using the measured boundary conditions of air according to the calculation formula, as shown in Equation (1).

$$T_{dp} = \frac{243.5 \times \ln\left(\frac{RH}{100} \exp \frac{17.67 \times T}{243.5+T}\right)}{17.67 - \ln\left(\frac{RH}{100} \exp \frac{17.67 \times T}{243.5+T}\right)} \quad (1)$$

By comparing two groups of measurements, it is possible to observe that the condensation capacity of a water wall depends on the temperature and relative humidity of the air. The first group includes measurements 5, 6, and 7, in which the air temperature was approximately equal (25.7 °C). The second group comprises measurements 8, 9, and 10, when the air temperature was approximately 27.3 °C, which is 1.6 °C higher than that of the first group. In the first case, the air temperature was 25.7 °C, and three measurements were carried out with relative humidity of 58.9% (RH_1), 74.0% (RH_2), and 83.8% (RH_3). In the second case, the air temperature was 27.3 °C and relative humidity 59.5% (RH_1), 75.7% (RH_2), and 83.6% (RH_3). The increase in relative humidity was similar in both cases. The plots (Figure 10) represent the courses of condensate weight created on the water film depending on time.

In the first case (Figure 10a), when relative humidity was 58.9%, the hourly gain of condensate was 21.99 g/h; at a relative humidity of 74.0%, the hourly increase in condensate was 148.81 g/h; and when relative humidity was 83.8%, the condensation performance was 236.73 g/h.

In the second case (Figure 10b), when the relative humidity was 59.5%, the hourly gain of condensate was 68.81 g/h; at a relative humidity of 75.7%, the hourly increase in condensate was 218.19 g/h; and when relative humidity was 85.6%, the condensation performance was 315.36 g/h.

The measurement results show that when the air temperature was 25.7 °C and relative humidity 58.9%, the condensation performance was 21.99 g/h. An increase in relative humidity of 15.1% represents a rise in condensate weight by 126.82 g, which is 6.8 times higher, whereas with an increase of 24.9%, condensate weight increases by 214.74 g, which is 10.8 times higher compared to when relative humidity is 58.9%.

When the air temperature increased by 1.6 °C to 27.3 °C, the dehumidification performance for a relative humidity of 59.5% was 68.81 g/h. An increase in relative humidity of 16.2% represents a rise in condensate weight by 149.38 g, which is 3.2 times higher, whereas with an increase of 26.1%, condensate weight increases by 246.55 g, which is 4.6 times higher compared to when relative humidity is 59.5%.

The results (Table 3) show a co-dependence of the dehumidification capacity of the water wall and boundary conditions. Even though the increase in relative humidity is comparable in both groups, the dehumidification capacity of the water wall is incomparable.

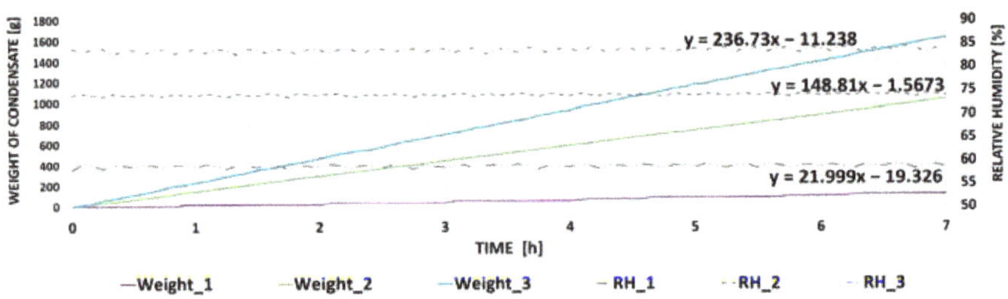

(a)

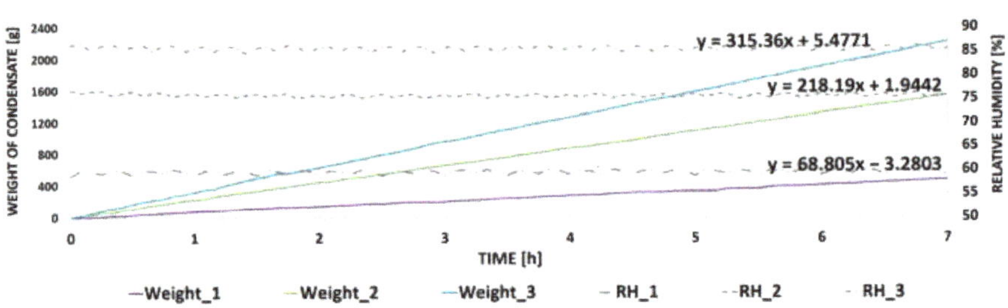

(b)

Figure 10. Courses of condensate weight increment: (**a**) First case—air temperature of 25.7 °C; (**b**) Second case—air temperature of 27.3 °C.

3.1. Analysis of Results Using Linear Dependence

Air humidity can be expressed in various ways. One of them is the humidity ratio. Based on the measured boundary conditions of air (relative humidity and air temperature), the humidity ratio was determined (Table 4). The plot in Figure 11 shows the linear dependence between humidity ratio (g/kg) and condensate rate (g/h).

From the graphical solution of linear dependence, Equation (2) was derived:

$$\text{CR} = 32.982 \times \chi - 376.64 \quad (2)$$

where CR is the condensation rate (g/h) and χ is the humidity ratio (g/kg). The correlation coefficient (R^2) of the equation is 0.9589. With the equation, it is possible to determine the water wall's condensation capacity for any humidity ratio.

In the next step, the condensation rate was calculated using Equation (2), and the percentage of deviation compared to the measured values was specified. This comparison, together with the boundary conditions, is shown in Table 4.

Table 2. Resulting values of measured parameters.

	Measurement														
	1	2	3	4	5	6	7	8	9	10	11	12	13	14	15
Temperature (°C)	22.2 ± 0.1	22.1 ± 0.1	24.0 ± 0.1	23.8 ± 0.1	25.7 ± 0.1	25.8 ± 0.1	25.7 ± 0.2	27.5 ± 0.1	27.2 ± 0.1	27.3 ± 0.2	28.9 ± 0.2	28.9 ± 0.1	30.7 ± 0.1	30.8 ± 0.1	32.5 ± 0.1
Max Temperature (°C)	22.4	22.4	24.2	24.0	26.0	26.0	26.1	27.7	27.5	27.7	29.3	29.3	31.0	31.1	32.8
Min Temperature (°C)	21.9	22.0	23.9	23.6	24.6	25.6	25.3	27.2	26.9	26.7	28.5	28.6	30.5	30.4	32.0
Relative humidity (%)	72.3 ± 1.4	82.7 ± 1.0	73.1 ± 1.4	83.7 ± 1.0	58.9 ± 1.4	74.0 ± 1.2	83.8 ± 1.3	59.5 ± 1.4	75.7 ± 1.0	85.6 ± 0.6	60.8 ± 0.9	70.9 ± 0.9	61.1 ± 1.0	71.6 ± 0.9	61.9 ± 0.9
Max Relative humidity (%)	75.1	84.5	75.5	85.5	62.7	76.2	86.3	62.1	78.0	87.1	62.5	72.7	63.0	73.5	64.5
Min Relative humidity (%)	70.2	80.2	71.0	81.3	56.8	72.4	81.5	57.3	73.6	84.1	59.2	69.4	59.6	69.9	60.4
Dew point temperature (°C)	17.02	19.04	18.88	20.89	17.07	20.78	22.78	18.86	22.56	24.65	20.56	23.1	22.36	25.04	24.23
Water temperature_upper (°C)	13.6 ± 0.02	14.0 ± 0.01	13.7 ± 0.02	14.0 ± 0.02	13.9 ± 0.02	13.9 ± 0.05	14.0 ± 0.02	13.8 ± 0.02	14.2 ± 0.01	14.3 ± 0.01	14.1 ± 0.01	14.2 ± 0.03	14.2 ± 0.01	14.3 ± 0.01	14.3 ± 0.01
Water temperature_down (°C)	14.1 ± 0.03	14.4 ± 0.01	14.4 ± 0.06	14.6 ± 0.02	14.6 ± 0.03	14.6 ± 0.03	14.8 ± 0.02	14.4 ± 0.01	15.0 ± 0.02	15.2 ± 0.02	14.8 ± 0.02	15.1 ± 0.04	15.1 ± 0.02	15.4 ± 0.02	15.3 ± 0.02
Duration of experiment (h)	15.0	15.0	12.0	15.0	12.0	12.0	14.0	12.0	7.0	7.0	15.0	14.0	15.0	7.0	7.0
Condensation rate (g/h)	46.02	104.29	88.95	173.42	21.99	148.81	236.73	68.81	218.19	315.36	118.62	219.92	177.4	298.17	240.4

Table 3. The resulting comparison and evaluation of two selected groups of measurements.

	Air Temperature (°C)	Relative Humidity (%)	Condensation Rate (g/h)	Increscent (-)
First group of measurements	25.7	58.9	21.99	reference value
		74.0	148.81	6.8 times higher than reference value
		83.8	236.73	10.8 times higher than reference value
Second group of measurements	27.3	59.5	68.81	reference value
		75.7	218.19	3.2 times higher than reference value
		85.6	315.36	4.6 times higher than reference value

Table 4. Comparison of measured and calculated condensation rate using humidity ratio.

	Measurement														
	1	2	3	4	5	6	7	8	9	10	11	12	13	14	15
Temperature (°C)	22.2	22.1	24.0	23.8	25.7	25.8	25.73	27.5	27.2	27.3	28.9	28.9	30.7	30.8	32.5
Relative humidity (%)	72.3	82.7	73.1	83.7	58.9	74.0	83.8	59.5	75.7	85.6	60.8	70.9	61.1	71.6	61.9
Humidity ratio (g/kg)	12.45	14.18	14.03	15.94	12.49	15.83	17.94	14.02	17.7	20.15	15.61	18.31	17.48	20.64	19.63
Measurement: CR (g/h)	46.02	104.29	88.95	173.42	21.99	148.81	236.73	68.81	218.19	315.36	118.62	219.92	177.4	298.17	240.4
Equation: CR (g/h)	33.99	91.04	86.10	149.09	35.31	145.47	215.06	85.77	207.14	287.95	138.21	227.26	199.89	304.11	270.80
Absolute difference	12.04	13.25	2.85	24.33	13.32	3.34	21.67	16.96	11.05	27.41	19.59	7.34	22.49	5.94	30.40
Percent deviation (%)	26.16	12.70	3.20	14.03	60.55	2.25	9.16	24.65	5.06	8.69	16.51	3.34	12.67	1.99	12.64

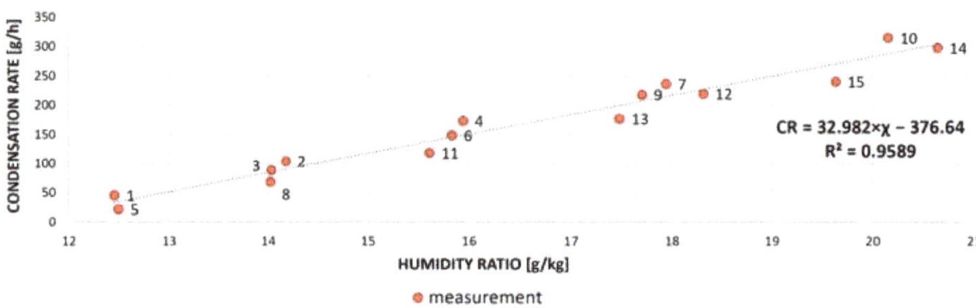

Figure 11. Linear dependence between humidity ratio and condensation rate.

The lowest percentage of deviation of the measured and calculated value of dehumidification performance was measurement number 14 (boundary conditions T = 30.7 °C, RH = 71.4%, χ = 20.64 g/kg), specifically 1.99%; in this case, the measured condensation rate was 298.17 g/h, and the calculated condensation rate according to Equation (2) was 304.11 g/h. The most significant difference between the measured and calculated condensation rate manifested in measurement number 5 (boundary conditions T = 25.7 °C, RH = 58.2%, χ = 12.49 g/kg). In this case, the measured condensation rate was 21.99 g/h, while using Equation (2), it was 35.31 g/h; this difference represents a percentage of deviation of 60.55%. The average percentage of deviation between all measured and calculated condensation rates was 14.24%.

3.2. Analysis of Results Using Nonlinear Surface Fitting

Due to fact that a high percentage of deviation between the measured and calculated condensation rate using a linear dependence analysis was discovered, for the next analysis, the dependence between condensation rate, air temperature, and relative humidity was determined using nonlinear surface fitting. The OriginPro software was chosen for this purpose. The 3D scatter plot (Figure 12a) shows the individual measurements in the spatial visualization.

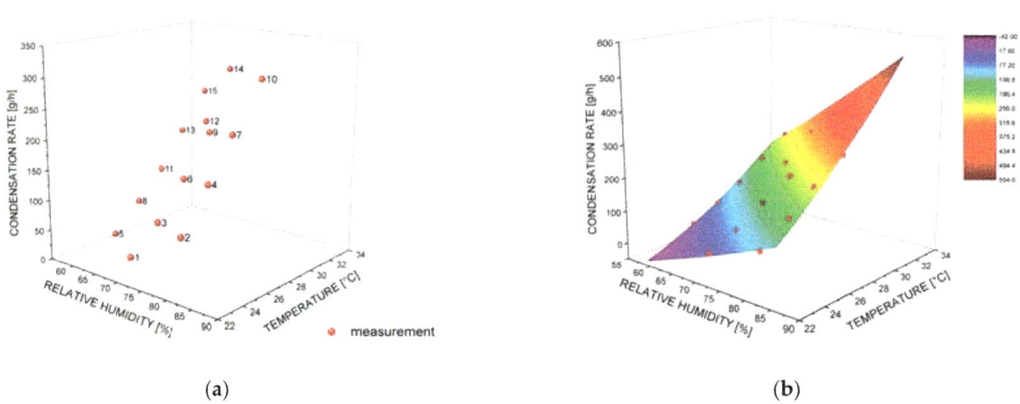

Figure 12. (**a**) 3D scatter plot—dependence between relative humidity, temperature, and condensation rate; (**b**) Nonlinear surface fitting tool predicted by OriginPro 9.0 software.

Using the two-dimensional polynomial function of the nonlinear surface fitting tool (Figure 12b), it was possible to determine the general Equation (3) of the water wall's condensation capacity as follows:

$$CR = z0 + a \times RH + b \times T + c \times RH^2 + d \times T^2 + f \times RH \times T \quad (3)$$

where CR is the condensation rate (g/h); z0, a, b, c, d, and f are constants (z0 = 662.003, a = −11.9448, b = −61.2747, c = 0.02751, d = 0.85944, and f = 0.64519); RH is the relative humidity (%); and T is the air temperature (°C). The coefficient of determination (R^2) of the equation is 0.99865. Equation (3) can be used for the calculation of the water wall's condensation capacity for any relative humidity and air temperature.

In the next step, the condensation rate was calculated using Equation (3), and the percentage of deviation compared to the measured values was specified. This comparison, together with the boundary conditions, is shown in Table 5.

Table 5. Comparison of condensate weight measured and calculated using nonlinear surface fitting.

	Measurement														
	1	2	3	4	5	6	7	8	9	10	11	12	13	14	15
Temperature (°C)	22.2	22.1	24.0	23.8	25.7	25.8	25.7	27.6	27.2	27.3	28.9	28.9	30.7	30.8	32.5
Relative humidity (%)	72.3	82.7	73.1	83.7	58.9	74.0	83.8	59.5	75.7	85.6	60.8	70.9	61.1	71.6	61.9
Measurement: CR (g/h)	46.02	104.29	88.95	173.42	21.99	148.81	236.73	68.81	218.19	315.36	118.62	219.92	177.40	298.17	240.40
Equation: CR (g/h)	42.07	107.88	92.19	169.27	23.52	150.71	237.63	68.35	214.15	315.90	118.08	222.98	174.84	297.67	241.88
Absolute difference	3.95	3.59	3.24	4.15	1.53	1.90	0.90	0.45	4.04	0.54	0.54	3.06	2.56	0.50	1.48
Percent deviation (%)	8.58	3.44	3.64	2.39	6.97	1.27	0.38	0.66	1.85	0.17	0.46	1.39	1.44	0.17	0.62

The lowest percentage of deviation of the measured and calculated value of dehumidification performance was found in measurement numbers 10 and 14 (boundary conditions no. 10: T = 27.3 °C, RH = 85.6%; no. 14: T = 30.8 °C, RH = 71.6%), specifically 0.17%. For measurement number 10, the measured condensation rate presented a value of 315.36 g/h, and the calculated rate according to Equation (3) was 315.90 g/h. For measurement number 14, the measured condensation rate presented a value of 298.17 g/h, and the rate calculated according to Equation (3) was 297.67 g/h. The most significant difference between the measured and calculated condensation rate manifested in measurement number 1 (boundary conditions: T = 22.2 °C and RH = 72.3%). In this case, the measured condensation rate was 46.02 g/h, while using Equation (3), it was 42.07 g/h; this difference represents a percentage of deviation of 8.58%. The average percentage of deviation between all measured and calculated condensation rates was 2.33%.

Air humidity, a significant indicator of indoor air quality, plays an important role, especially in the context of thermal comfort, where it is shown that, at the same temperature but different humidity, it is possible to feel different levels of comfort (with a high water vapor content, the environment is perceived as humid, and with a low water vapor content, as dry) [48]. The level of indoor humidity depends not only on the content of water vapor in outdoor air but also on various sources of indoor humidity and people's activities. Decorative water elements (fountains, water walls, pools) are perceived as humidifiers [1]. However, the study [49] shows that through the substitution of water with liquid desiccant, the vase or sphere with the falling film on the surface can act as a moisture receptacle that removes moisture at high humidity and releases moisture at low humidity. Thus, decorative elements can also be used to adjust humidity level.

The presented paper monitors the condensation rate on a falling water film under different boundary conditions. The values of measurements were chosen outside the area of thermal comfort (according to the ASHRAE standard [48]), especially with regard to the fact that in such a case, it is necessary to adjust the air (cooling and dehumidification) to achieve the required level of thermal comfort. Obtaining negative results using the presented Equations (2) and (3) means that dehumidification will not take place; for the respective boundary conditions, Equations (2) and (3) cannot be used (Figure 13).

Tepmerature [°C]	22	22	22	22	22	22	22	22	22	22	22	22	22	22	22
Relative humidity [%]	30	35	40	45	50	55	60	65	70	75	80	85	90	95	100
Condensation rate [g/h]	no	no	no	no	no	no	no	no	22.19	53.38	85.94	119.89	155.20	191.90	229.97
Tepmerature [°C]	23	23	23	23	23	23	23	23	23	23	23	23	23	23	23
Relative humidity [%]	30	35	40	45	50	55	60	65	70	75	80	85	90	95	100
Condensation rate [g/h]	no	no	no	no	no	no	no	11.71	44.75	79.17	114.96	152.13	190.67	230.59	271.89
Temperature [°C]	24	24	24	24	24	24	24	24	24	24	24	24	24	24	24
Relative humidity [%]	30	35	40	45	50	55	60	65	70	75	80	85	90	95	100
Condensation rate [g/h]	no	no	no	no	no	no	no	32.76	69.03	106.67	145.69	186.09	227.86	271.00	315.52
Temperature [°C]	25	25	25	25	25	25	25	25	25	25	25	25	25	25	25
Relative humidity [%]	30	35	40	45	50	55	60	65	70	75	80	85	90	95	100
Condensation rate [g/h]	no	no	no	no	no	no	17.42	55.54	95.03	135.90	178.15	221.77	266.76	313.13	360.88
Temperature [°C]	26	26	26	26	26	26	26	26	26	26	26	26	26	26	26
Relative humidity [%]	30	35	40	45	50	55	60	65	70	75	80	85	90	95	100
Condensation rate [g/h]	no	no	no	no	no	no	38.69	80.03	122.75	166.85	212.32	259.17	307.39	356.98	407.96
Temperature [°C]	27	27	27	27	27	27	27	27	27	27	27	27	27	27	27
Relative humidity [%]	30	35	40	45	50	55	60	65	70	75	80	85	90	95	100
Condensation rate [g/h]	no	no	no	no	no	18.48	61.67	106.25	152.19	199.51	248.21	298.28	349.73	402.55	456.75
Temperature [°C]	28	28	28	28	28	28	28	28	28	28	28	28	28	28	28
Relative humidity [%]	30	35	40	45	50	55	60	65	70	75	80	85	90	95	100
Condensation rate [g/h]	no	no	no	no	no	39.96	86.38	134.18	183.35	233.90	285.82	339.12	393.79	449.84	507.27
Temperature [°C]	29	29	29	29	29	29	29	29	29	29	29	29	29	29	29
Relative humidity [%]	30	35	40	45	50	55	60	65	70	75	80	85	90	95	100
Condensation rate [g/h]	no	no	no	no	14.89	63.16	112.81	163.83	216.23	270.00	325.15	381.67	439.57	498.85	559.50
Temperature [°C]	30	30	30	30	30	30	30	30	30	30	30	30	30	30	30
Relative humidity [%]	30	35	40	45	50	55	60	65	70	75	80	85	90	95	100
Condensation rate [g/h]	no	no	no	no	36.58	88.08	140.95	195.20	250.82	307.82	366.20	425.95	487.07	549.57	613.45
Temperature [°C]	31	31	31	31	31	31	31	31	31	31	31	31	31	31	31
Relative humidity [%]	30	35	40	45	50	55	60	65	70	75	80	85	90	95	100
Condensation rate [g/h]	no	no	no	6.64	59.99	114.71	170.81	228.29	287.14	347.36	408.96	471.94	536.29	602.02	669.12
Temperature [°C]	32	32	32	32	32	32	32	32	32	32	32	32	32	32	32
Relative humidity [%]	30	35	40	45	50	55	60	65	70	75	80	85	90	95	100
Condensation rate [g/h]	no	no	no	28.55	85.12	143.07	202.39	263.09	325.17	388.62	453.45	519.65	587.23	656.18	726.51
Temperature [°C]	33	33	33	33	33	33	33	33	33	33	33	33	33	33	33
Relative humidity [%]	30	35	40	45	50	55	60	65	70	75	80	85	90	95	100
Condensation rate [g/h]	no	no	no	52.17	111.97	173.14	235.69	299.62	364.92	431.60	499.65	569.08	639.88	712.06	785.62
Temperature [°C]	34	34	34	34	34	34	34	34	34	34	34	34	34	34	34
Relative humidity [%]	30	35	40	45	50	55	60	65	70	75	80	85	90	95	100
Condensation rate [g/h]	no	no	15.86	77.51	140.54	204.94	270.71	337.87	406.39	476.30	547.57	620.23	694.26	769.66	846.44
Temperature [°C]	35	35	35	35	35	35	35	35	35	35	35	35	35	35	35
Relative humidity [%]	30	35	40	45	50	55	60	65	70	75	80	85	90	95	100
Condensation rate [g/h]	no	no	39.69	104.57	170.82	238.45	307.45	377.83	449.58	522.71	597.22	673.10	750.35	828.98	908.99

no working | CR ≤ 100 g/h | 100 < CR ≤ 200 g/h | 200 < CR ≤ 300 g/h | 300 < CR ≤ 400 g/h | 400 < CR ≤ 500 g/h | CR > 500 g/h

Figure 13. Condensation rate on a falling water film with an area of 1 m^2 for the selected temperature and relative humidity using Equation (3).

Currently, various air-conditioning systems are most often used to achieve required humidity levels. Condensate dehumidifiers and desiccant dehumidifiers are used most often [50]. On the market, it is possible to find several commercial systems, from the smallest designed for small rooms, whose dehumidification capacity is about 0.3 L per 24 h, through medium mobile dehumidifiers designed for individual rooms at home with an average dehumidification capacity of about 8–30 L per 24 h (at a temperature of 30 °C and relative humidity of 80%) and to large complex air-handling units with a built-in system for adjusting temperature and humidity, which are suitable for buildings with controlled ventilation. At an air temperature of 30 °C and relative humidity of 80%, the dehumidification capacity of a water wall with an effective area of 1 m^2, according to Equation (3), is 366.2 g/h. This is approximately 8.8 L per 24 h, and this dehumidification capacity is comparable to that of a medium mobile dehumidifier.

Condensation on thin falling liquid films that fall on solid surfaces is a common and very important process for technical applications and is wildly used in air-conditioning systems [37]. The results of this paper show that this process can also be used in the case of water walls with a controlled water temperature. Conventional air-conditioning systems are closed systems with controlled inlet air velocity and direction, compared to water walls where dehumidification occurs directly in the room where they are placed and without airflow control.

4. Conclusions

This paper shows the possibility of the use of water walls as dehumidifiers, although this element is built in interior spaces mainly for its aesthetic function. The subject of the research was a water wall with controlled water temperature and effective area of falling water film of 1 m^2. The main conclusions are as follows:

- Testing in laboratory conditions in the climate chamber allowed the definition of water walls with controlled water temperature as a device with a certain level of performance. The measurement of increments of condensate using the laboratory scale ensured the accurate definition of changes in time.
- The results showed that dehumidification capacity is a dynamic process that strongly depends on the boundary conditions of the air.
- The water wall dehumidification capacity varied in the range of 21.99–315.36 g/h for the tested measurements. This represents a difference between the lowest and highest performance of 293.37 g/h.
- Equation (3) determined by nonlinear surface fitting shows a higher coefficient of determination (R^2 = 0.99865) and a lower average percentage of deviation between measured and calculated values (2.33%) compared to Equation (2) generated by linear dependence (R^2 = 0.9589; average percentage of deviation = 14.24%).
- The achieved results expand our knowledge regarding the use of water elements to adjust the parameters of indoor air and their ability to connect with convectional air treatment systems with the aim to reduce energy consumption.

A water wall with a controlled water temperature was tested in laboratory conditions in a climate chamber in order to determine its dehumidification capacity regardless of the surrounding materials or room volume. However, the real effects of water walls in interior spaces will depend on the sorption capacity of the elements and the surrounding materials in the interior of buildings. Because of this, it is necessary to develop this topic and to examine the dehumidification capacity of water walls in real conditions of sorption-active surfaces.

Author Contributions: Conceptualization, K.C. and F.V.; methodology, K.C.; software, M.V.; validation, K.C., F.V. and M.V.; formal analysis, K.C.; investigation, K.C.; resources, K.C.; data curation, K.C.; writing—original draft preparation, K.C.; writing—review and editing, Z.V. and M.V.; visualization, K.C.; supervision, Z.V.; project administration, Z.V.; funding acquisition, Z.V. All authors have read and agreed to the published version of the manuscript.

Funding: This research received no external funding.

Institutional Review Board Statement: Not applicable.

Informed Consent Statement: Not applicable.

Data Availability Statement: All data are presented in this article in the form of figures and tables.

Acknowledgments: The authors are extremely grateful to the support in this undertaking of the Ministry for Education of the Slovak Republic with VEGA 1/0217/19 "Research of Hybrid Blue and Green Infrastructure as Active Elements of a Sponge City", a project run by the Slovak Research and Development Agency APVV-18-0360 "Active hybrid infrastructure closer to a sponge city" and a project run by the Operational Programme Research, Development and Education of The Ministry of Education, Youth and Sports of the Czech Republic CZ.02.2.69/0.0/0.0/18_053/0016962 "International mobility of researchers at the Brno University of Technology II".

Conflicts of Interest: The authors declare no conflict of interest.

References

1. Wu, T.; Lei, C. A review of research and development on water wall for building applications. *Energy Build.* **2016**, *112*, 198–208. [CrossRef]
2. Wu, T.; Lei, C. Thermal modelling and experimental validation of a semi-transparent water wall system for Sydney climate. *Sol. Energy* **2016**, *136*, 533–546. [CrossRef]

3. Santamaria, M.B.; Gonzalo, F.A.; Pinette, D.; Lauret Aguirregabiria, B.; Hernandez Ramos, J.A. Industrialization and Thermal Performance of a New Unitized Water Flow Glazing Façade. *Sustainability* **2020**, *12*, 7564. [CrossRef]
4. Wang, Y.; Liew, J.Y.R.; Lee, S.C. A novel multi-functional water façade system for energy saving and blast resisting. *Mater. Des.* **2016**, *106*, 98–111. [CrossRef]
5. Rathnayake, U.; Lau, D.; Chow, C.L. Review on Energy and Fire Performance of Water Wall Systems as a Green Building Façade. *Sustainability* **2020**, *12*, 8713. [CrossRef]
6. Wilson, E.O. *Biophilia: The Human Bond with Other Species*; Harvard University Press: Cambridge, UK, 1984.
7. Kellert, S.R.; Heerwagen, J.; Mador, M. *Biophilic Design: The Theory, Science and Practice of Bringing Buildings to Life*; John Wiley & Sons: Hoboken, NJ, USA, 2011.
8. Browning, W.; Ryan, C.; Clancy, J. *14 Patterns of Biophilic Design*; Terrapin Bright Green, LLC: New York, NY, USA, 2014; Available online: http://www.terrapinbrightgreen.com/wp-content/uploads/2014/09/14-Patterns-of-Biophilic-Design-Terrapin-2014p.pdf (accessed on 19 April 2021).
9. Charoenkit, S.; Yiemwattana, S. Living walls and their contribution to improved thermal comfort and carbon emission reduction: A review. *Build. Environ.* **2016**, *105*, 82–94. [CrossRef]
10. Wang, C.; Er, S.-S.; Abdul-Rahman, H. Indoor vertical greenery system in urban tropics. *Indoor Built Environ.* **2016**, *25*, 340–356. [CrossRef]
11. Franco, A.; Fernández-Cañero, R.; Pérez-Urrestarazu, L.; Valera, D.L. Wind tunnel analysis of artificial substrates used in active living walls for indoor environment conditioning in Mediterranean buildings. *Build. Environ.* **2012**, *51*, 370–378. [CrossRef]
12. Gunawardena, K.; Steemers, K. Living walls in indoor environments. *Build. Environ.* **2019**, *148*, 478–487. [CrossRef]
13. Moya, A.T.; van den Dobbelsteen, A.; Ottele, M.; Bluyssen, M.P. A review of green systems within the indoor environment. *Indoor Built Environ.* **2019**, *28*, 298–309. [CrossRef]
14. Pérez-Urrestarazu, L.; Fernández-Cañero, R.; Franco, A.; Egea, G. Influence of an active living wall on indoor temperature and humidity conditions. *Ecol. Eng.* **2016**, *90*, 120–124. [CrossRef]
15. Tudiwer, D.; Korjenic, A. The effect of an indoor living wall system on humidity, mould spores and CO2-concentration. *Energy Build.* **2017**, *1461*, 73–86. [CrossRef]
16. Poórová, Z.; Turcovská, A.; Kapalo, P.; Vranayová, Z. The Effect of Green Walls on Humidity, Air Temperature, Co2 and Well-Being of People. *Environ. Sci. Proc.* **2020**, *2*, 2056. [CrossRef]
17. Horr, A.Y.; Arif, M.; Kaushik, A.; Mazroei, A.; Katafygiotou, M.; Elsarrag, E. Occupant productivity and office indoor environment quality: A review of the literature. *Build. Environ.* **2016**, *105*, 369–389. [CrossRef]
18. Lin, F.G. *Human Responses to Water Elements in Interior Environments: A Culture and Gender Comparison. The Handbook of Interior Design: Chapter*; John Wiley & Sons, Ltd.: Hoboken, NJ, USA, 2015; pp. 293–309. [CrossRef]
19. Alvarsson, J.J.; Wien, S.; Nilsson, M.E. Stress recovery during exposure to nature sound and environmental noise. *Int. J. Environ. Res. Public Health* **2010**, *7*, 1036–1046. [CrossRef]
20. Yang, W.; Moon, J.H.; Kim, M.-J. Perceptual assessment of indoor water sounds over environmental noise through windows. *Appl. Acoust.* **2018**, *135*, 60–69. [CrossRef]
21. Chen, W.; Liu, S.; Lin, J. Analysis on the passive evaporative cooling wall constructed of porous ceramic pipes with water sucking ability. *Energy Build.* **2015**, *86*, 541–549. [CrossRef]
22. Moustafa, M.A.; Aripin, S. CFD evaluation of the pottery water wall in a hot arid climate of Luxor, Egypt. *J. Green Build.* **2014**, *9*, 175–189. [CrossRef]
23. Tewari, P.; Mathur, S.; Mathur, J. Thermal performance prediction of office buildings using direct evaporative cooling systems in the composite climate of India. *Build. Environ.* **2019**, *157*, 64–78. [CrossRef]
24. Moosavi, L.; Mahyuddin, N.; Ghafar, N. Atrium cooling performance in a low energy office building in the Tropics, a field study. *Build. Environ.* **2015**, *94*, 384–394. [CrossRef]
25. Yin, Y.L.; Wang, R.Z.; Zhai, X.Q.; Ishugah, T.F. Experimental investigation on the heat transfer performance and water condensation phenomenon of radiant cooling panels. *Build. Environ.* **2014**, *71*, 15–23. [CrossRef]
26. Le Dréau, J.; Heiselberg, P. Sensitivity analysis of the thermal performance of radiant and convective terminals for cooling buildings. *Energy Build.* **2014**, *82*, 482–491. [CrossRef]
27. Haupt, T.E.; Heffernan, T.R.; Kazmierczak, J.J.; Nehls-Lowe, H.; Rheineck, B.; Powell, C.; Leonhardt, K.K.; Chitnis, S.A.; Davis, P.J. An Outbreak of Legionnaires Disease Associated with a Decorative Water Wall Fountain in a Hospital. *Infect. Control Hosp. Epidemiol.* **2012**, *33*, 185–191. [CrossRef] [PubMed]
28. Mitterer, C. *Cooling and Dehumidifying Indoor Environments Using the Chilled Water Wall*; IBP-FRAUNHOFER: Valley, Germany; Available online: https://www.ibp.fraunhofer.de/content/dam/ibp/en/documents/Information-material/Departments/Hygrothermics/Produktblaetter/IBP_255_PB_Klimabrunnen_neu_en_rz_web.pdf (accessed on 19 April 2021).
29. Fang, Y.-M.; Lee, J.M.C.; Mitterer, C.; Kunzel, M.H. The performance of heat pump integrated with chilled water wall for air conditioning and domestic hot water. *IOP Conf. Ser. Earth Environ. Sci.* **2019**, *294*, 012032. [CrossRef]
30. Oliveira, R.; Figueiredo, A.; Vicente, R.; Almeida, M.S.F.R. Impact of unoccupied flats on the thermal discomfort and energy demand: Case of a multi-residential building. *Energy Build.* **2020**, *209*, 109704. [CrossRef]
31. Juras, P.; Durica, P.; Jandacka, J. Parametric Study: Impact of Selected Factors on the Energy Demand of a Family House. *Iop Conf. Ser. Earth Environ. Sci.* **2019**, *290*, 012096. [CrossRef]

32. Dolnikova, E.; Katunsky, D.; Vertal, M.; Zozulak, M. Influence of Roof Windows Area Changes on the Classroom Indoor Climate in the Attic Space: A Case Study. *Sustainability* **2020**, *12*, 5046. [CrossRef]
33. Qi, R.; Dong, C.; Zhang, L.-Z. A review of liquid desiccant air dehumidification: From system to material manipulations. *Energy Build.* **2020**, *215*, 109897. [CrossRef]
34. Cellura, M.; Guarino, F.; Longo, S.; Tumminia, G. Climate change and the building sector: Modelling and energy implications to an office building in southern Europe. *Energy Sustain. Dev.* **2018**, *45*, 46–65. [CrossRef]
35. Isaac, M.; van Vuuren, P.D. Modeling global residential sector energy demand for heating and air conditioning in the context of climate change. *Energy Policy* **2009**, *37*, 507–521. [CrossRef]
36. Vertal', M.; Zozulák, M.; Vašková, A.; Korjenic, A. Hygrothermal initial condition for simulation process of green building construction. *Energy Build.* **2018**, *167*, 166–176. [CrossRef]
37. Sosnowski, P.; Petronio, A.; Armenio, V. Numerical model for thin liquid film with evaporation and condensation on solid surfaces in systems with conjugated heat transfer. *Int. J. Heat Mass Transf.* **2013**, *66*, 382–395. [CrossRef]
38. Kim, H.; Chen, L.; Lee, J. Characteristics of vertically falling liquid film affected by cross gas flow in a narrow flat channel. *Nucl. Eng. Des.* **2020**, *357*, 110393. [CrossRef]
39. Åkesjö, A.; Gourdon, M.; Vamling, L.; Innings, F.; Sasic, S. Experimental and numerical study of heat transfer in a large-scale vertical falling film pilot unit. *Int. J. Heat Mass Transf.* **2018**, *125*, 53–65. [CrossRef]
40. Lu, C.; Jiang, S.-Y.; Duan, R.-Q. Wave Characteristics of Falling Film on Inclination Plate at Moderate Reynolds Number. *Sci. Technol. Nucl. Install.* **2016**, *2016*, 1–7. [CrossRef]
41. Bird, B.R.; Stewart, E.W.; Lightfoot, N.E. *Transport Phenomena*, 2nd ed.; John Wiley and Sons: New York, NY, USA, 2007.
42. Zhou, D.W.; Gambaryan-Roisman, T.; Stephan, P. Measurement of water falling film thickness to flat plate using confocal chromatic sensoring technique. *Exp. Therm. Fluid Sci.* **2009**, *33*, 273–283. [CrossRef]
43. Nasr, A.; Al-Ghamdi, S.A. Numerical study of evaporation of falling liquid film on one of two vertical plates covered with a thin porous layer by free convection. *Int. J. Therm. Sci.* **2017**, *112*, 335–344. [CrossRef]
44. Gonda, A.; Lancereau, P.; Bandelier, P.; Luo, L.; Fan, Y.; Benezech, S. Water falling film evaporation on a corrugated plate. *Int. J. Therm. Sci.* **2014**, *81*, 29–37. [CrossRef]
45. Čákyová, K.; Vranay, F. *Utility Model Number 8710: Water Wall Collection Container with Overflow Edge (Zberná Nádoba Vodnej Steny s Prepadovou Hranou)*; Industrial Property Office of the Slovak Republic: Banská Bystrica, Slovak, 2019. Available online: https://wbr.indprop.gov.sk/WebRegistre/UzitkovyVzor/Detail/50129-2017 (accessed on 19 April 2021).
46. Han, J.; Guo, H. Development of a method for condensation rate measurement on flat surfaces. *Inf. Process. Agric.* **2018**, *5*, 490–497. [CrossRef]
47. Čákyová, K.; Vranay, F.; Kušnír, M. Impact of flow rate to water film thickness of water wall. In *Advances and Trends in Engieneering Sciences and Technologies III*; CRC Press: London, UK, 2019; pp. 337–342. [CrossRef]
48. ASHRAE. Thermal comfort. In *Ashrae Handbook Fundamentals*; American Society of Heating, Refrigerating and Air-Conditioning Engineering, Inc.: Atlanta, GA, USA, 2001.
49. Jianbo, C.; Xue, X.; Minglu, Q.; Xing, H.; Youli, K. Experimental analysis on dehumidification performance of an indoor passive falling film liquid desiccant moisture receptacle. *Energy Build.* **2016**, *125*, 161–170. [CrossRef]
50. Ge, F.; Wang, C. Exergy analysis of dehumidification systems: A comparison between the condensing dehumidification and the desiccant wheel dehumidification. *Energy Convers. Manag.* **2020**, *224*, 113343. [CrossRef]

sustainability

MDPI

Article

Performance of Conventional and Innovative Single U-Tube Pipe Configuration in Vertical Ground Heat Exchanger (VGHE)

Adel Eswiasi [1] and Phalguni Mukhopadhyaya [2,*]

[1] Mechanical Engineering, University of Victoria, Victoria, BC V8W 2Y2, Canada; aeswiasi@uvic.ca
[2] Civil Engineering, University of Victoria, Victoria, BC V8W 2Y2, Canada
* Correspondence: phalguni@uvic.ca

Abstract: A ground source heat pump system (GSHP) with a ground heat exchanger (GHE) is a renewable and green technology used for heating and cooling residential and commercial buildings. An innovative U-Tube pipe configuration is suggested to enhance the heat transfer rate in the vertical ground heat exchanger (VGHE). Laboratory experiments are conducted to compare the thermal efficiency of VGHEs with two different pipe configurations: (1) an innovative U-Tube pipe configuration (single U-Tube with two outer fins) and (2) a single U-Tube. The results show that the difference between the inlet and outlet temperatures for the innovative U-Tube pipe configuration was 0.7 °C after 60 h, while it was 0.4 °C for the single U-Tube after the same amount of time. The borehole thermal resistance for the innovative U-Tube pipe configuration was 0.680 m·K/W, which is 29.22% lower than that of the single U-Tube. The heat exchange rate in the innovative U-Tube pipe configuration is increased by 57.95% compared to the conventional single U-Tube. Measured ground temperatures indicate that compared to single U-Tube pipe configuration, the innovative U-Tube pipe configuration has superior heat transfer performance. Based on the experimental results presented in this paper, it was concluded that increasing the surface area significantly by introducing external fins to the U-Tube enhances the heat transfer rate, resulting in increased thermal efficiency of the VGHE.

Keywords: vertical ground heat exchanger; external fin; thermal response test; effective ground thermal conductivity

Citation: Eswiasi, A.; Mukhopadhyaya, P. Performance of Conventional and Innovative Single U-Tube Pipe Configuration in Vertical Ground Heat Exchanger (VGHE). *Sustainability* **2021**, *13*, 6384. https://doi.org/10.3390/su13116384

Academic Editor: Mostafa Ghasemi Baboli

Received: 30 April 2021
Accepted: 31 May 2021
Published: 4 June 2021

1. Introduction and Background

A ground source heat pump system (GSHP) is an ecofriendly technology utilized for heating and cooling houses and commercial buildings. GSHP technology exploits the constant ground temperature over the year to extract heat from buildings and transfer it into the ground in summer, as well as to extract heat from the ground and transfer it into the buildings in winter [1,2]. The GSHP system is constructed by connecting a heat pump with a vertical ground heat exchanger (VGHE). Since the second half of the 1990s, Thermal Response Tests (TRTs) have been used to measure ground thermal properties (borehole thermal resistance and effective ground thermal conductivity) and characterize the efficiency of VGHEs. Researchers have suggested different approaches to measure ground thermal properties in the field, and a number of mobile experimental apparatuses have been built in different countries to carry out these measurements [3]. The initial capital cost of installing the GSHP systems is a significant consideration in VGHE design. Hence, there are opportunities for designers to do more work in this area and suggest new pipe configurations to increase the heat transfer rate in the VGHEs and thus decrease the depth and cost of borehole installation. During the past few decades, several researchers conducted studies to improve the thermal performance of VGHEs by decreasing the thermal resistance between the borehole wall and the ground as much as possible. These studies were based on either numerical or experimental models, the latter includes a small-scale apparatus in the laboratory or a full-scale field investigation.

Gu and O'Neal (1998) built a small-scale single U-Tube VGHE in the laboratory to estimate the effects of backfills (bentonite/masonry sand and bentonite/copper powder) on the heat transfer performance of a single U-Tube VGHE. The single U-Tube copper pipe (inner diameter, 4.8 mm; outer diameter, 6.4 mm; length, 1.2 m) was inserted in a metal tank. The obtained results showed that backfill materials have an influence on the thermal performance of single U-Tube VGHEs [4].

To determine the effects of another factor, Pahud and Matthey (2001) conducted TRTs for the double U-Tube VGHEs, with and without spacers, to estimate the influence of these spacers on the borehole thermal resistance. The borehole thermal resistances of double U-Tube VGHEs with and without spacers were 0.141 m·K/W and 0.143 m·K/W, respectively [5]. Zeng et al. (2003) suggested an analytical solution considering the influence of fluid axial convective heat transfer on pipe configurations of VGHEs. Two different configurations, single and double U-Tube, of VGHEs were examined by using this method. The results showed that the borehole thermal resistance of double U-Tube VGHE was lower than the single U-Tube VGHE [6].

Only one study conducted by Esen and Inalli (2009) showed the effects of three different borehole depths (30 m, 60 m, and 90 m) on the thermal performance of VGHEs. The results showed that the 90 m depth borehole heat exchanger had stronger performance than those with the depths of 60 m and 30 m. It was also noticed that the thermal resistance values for the single U-Tube VGHE with depths of 60 m and 90 m were 0.05 and 0.03 m·K/W, respectively [7].

To determine the effects of different pipe configurations on the thermal performance of VGHE, Acuña and Palm (2010) proposed a new pipe configuration, coaxial (pipe-in-pipe) VGHE, and it compared with conventional single U-Tube. The outer pipe was in direct contact with the surrounding bedrock. The results showed that the heat transfer performance of the coaxial VGHE was superior to the conventional single U-Tube VGHE [8]. Another field study was conducted by Lee et al. (2011) to estimate the thermal efficiency of two different pipe configurations, the common U-Tube type and a new three-pipe type GHE. The thermal efficiency of the new three-pipe type configuration was higher than that of the conventional single U-Tube type [9]. In a study that set out to determine the effects of three different types—coaxial, double U-Tube, and U-Tube VGHEs—in Oklahoma City, OK, USA, Beier and Ewbank (2012) found that the best option to reduce the thermal resistance of the borehole was the double U-Tube, followed by the single U-Tube, and finally the coaxial type [10]. In the same year, a field study was performed by Desmedt et al. (2012) to estimate the influence of two different pipe configurations (single U-pipe and double U-pipe) on the efficiency of the VGHEs. The thermal resistance of the double U-Tube VGHE was 0.162 m·K/W, 52% lower than the single U-Tube VGHE [11]. To improve the thermal performance of VGHEs, a numerical study was investigated by Haddada and Miyara (2014) to estimate the impacts of pipe numbers inside the borehole on the performance of VGHEs. The four different pipe configurations examined were U-Tube, multi-tube, three-tube, and four-tube. The heat exchange rate increased between the boreholes and the ground due to an increase in the number of inlet tubes inside the borehole [12].

Kramer and Basu (2014) built an experimental apparatus in the laboratory to study the effect of thermal loading on load displacement behavior of the model geothermal pile. The polyvinylchloride (PVC) U-Tube pipe was installed in a sand tank (length, 1.83 m; depth, 1.83 m; width, 1.83 m). The U-Tube pipe consisted of two PVC pipes that had an inner diameter of 12.4 mm with a length of 1.22 m [13]. Shirazi and Bernier (2014) established a small-scale experimental apparatus with a borehole Plexiglas pipe (length, 1.23 m) inserted at the center from the top of the sand tank (length, 1.35 m; diameter, 1.4 m). The TRT was 73 h for the heat injection and 5 days to let the sand return to its initial ground temperature. It was reported that the borehole thermal resistance for the single U-Tube VGHE was 0.61 m·K/W [14]. Another small scale VGHE study was conducted by Erol and François (2014) to estimate the effects of backfill materials on the thermal performance of the VGHE. Two parallel pipes were inserted in the 1 m length of the VGHE in the sand tank of 1 m^3.

Two heat pumps were used to pump the water to pipes at 12 °C, 15 °C inlet and outlet of water temperatures, respectively. The initial temperature of the sand was 20 °C [15]. One year later, Cimmino and Bernier (2015) built the experimental setup consisting of a 400 mm tall borehole inserted in center and top of the sand tank (length, 1.35 m; diameter, 1.4 m). The single U-Tube copper pipe was inserted in the borehole. The objective of this study was to measure the borehole wall temperature at various depths and different times during the TRT. The results showed that the borehole wall temperature increased with an increase in the duration of the TRT [16].

Liu et al. (2015) suggested a new VGHE design with three inlet pipes and one outlet (3I-type) and compared its performance to the single and double U-Tube VGHEs. Experimental results showed that the thermal resistance of the 3I-type was 31% and 15.8% lower than the single and double U-Tubes, respectively [17]. In an analysis of the influences of the different pipe material properties (copper and high density polyethylene pipes) on the performance of VGHEs, Ramadan (2016) found that the performance trend remains the same for both copper and high density polyethylene pipes in a ground heat exchanger [18]. Another field was conducted by Chang and Kim (2016) to estimate the influence of two different pipe configurations, single U-Tube and double U-Tube, on the thermal performance of VGHEs. The thermal resistance of the single and double U-Tubes was 0.130 m·K/W and 0.081 m·K/W, respectively [19]. In a numerical study investigated by Luo et al. (2016), the thermal efficiency of four different pipe configurations (double-U, triple-U, double-W, and spiral) was estimated. The triple U-Tube had a higher thermal efficiency, followed by the double-W, the spiral type, and the double U-Tube. The results also indicated that the double-W type had the lowest economic performance, followed by the spiral type, the double-U type, and the triple-U type [20]. Two years later, another numerical study was investigated by Serageldin et al. (2018) to improve the performance of VGHE by suggesting an innovative U-Tube pipe configuration (an oval U-Tube), and it compared a conventional single U-Tube. The results showed that the thermal resistance of the oval U-Tube was 0.125 m·K/W, and 15.8% lower than the conventional single U-Tube [21]. A field study was conducted by Bae et al. (2019) to estimate the thermal performance for four different pipe types (high-density polyethylene (HDPE) type, HDPE-nano type, spiral fin type, and coaxial type). The results showed that the best option to reduce the thermal resistance of the borehole was the spiral fin type (0.181 m·K/W), followed by the HDPE-nano type (0.181 m·K/W), HDPE type (0.183 m·K/W), and a coaxial type (0.306 m·K/W) [22].

A recent laboratory study was conducted by Li et al. (2018) to estimate the effects of ground stratification on the performance of GHEs. Two U-Tube copper GHEs were installed in a sand tank (depth, 6.25 m; length, 1.5 m; width, 1 m), and the sand tank was filled with sand and clay. The two U-Tube pipe consisted of two copper pipes that had an inner diameter of 5 mm and diameter outer of 5.5 mm with a length of 6.25 m. The results showed that an increase in the ground heat injection rates leads to a significant increase in the effects of ground stratification [23]. One year later, Liang et al. (2019) constructed an experimental setup with a small-scale spiral-tube copper pipe VGHE inserted in the metal container (diameter, 0.8 m; height, 1.1 m) setup to investigate the influences of two parameters (volume flow rate and backfill materials) on the thermal performance of a small-scale spiral-tube VGHE. The results showed that the heat exchange rate in the spiral-tube VGHE in blend added to the sand with different Reynolds numbers (3000–8000) enhanced by 20–31% compared to the native sand [24].

To sum up, the previous paragraphs present a comprehensive review of the efficiency of ground heat exchangers (GHE) in heat pump systems. Increasing the efficiency of the GHE may lead to decreased borehole depths and installation costs. Hence, the identification and optimization of design and operation parameters that can reduce the borehole depth are among the most important design challenges. To deal with these challenges, various researchers have conducted several studies to improve the heat transfer efficiency of GHEs. However, there are still opportunities for engineers to suggest new pipe configurations that can increase the surface area of the pipe and heat transfer rate in GHEs. One of the most attractive solutions to increase the heat transfer rate in VGHEs is to increase the surface area of the pipe configuration, which could lead to higher heat exchange capacity of VGHEs. It is well known that heat transfer from a surface can be increased by attaching external fins (i.e., extended surfaces). However, research activities reported in the literature do not provide any information about the performance of GHEs with external fins. Therefore, this research initiative investigates the impacts of external fins on the heat transfer capacity of VGHEs.

In this paper, an innovative U-Tube pipe configuration (conventional single U-Tube with two outer fins) is proposed to improve the thermal performance of VGHE by increasing the surface area for a higher heat transfer rate. Fins or extended surfaces are used to enhance heat transfer by conduction through the solid and also by convection from the solid boundaries. Fins of many shapes such as rectangular, trapezoidal, concave, etc. could be used to enhance the heat transfer rate. The concave shape has a higher heat transfer rate than the rectangular or trapezoidal shape, and the rectangular shape is more efficient in transferring heat than the trapezoidal shape [25]. In this study, the fins were planned to be designed in a concave shape, but then it was decided to change them to a trapezoidal shape because the tools available in the workshop were not suitable for producing fins in a concave shape. Another secondary reason to select the trapezoidal shape is its small tip at the end of the fin. This means it can be a good option to increase the fin length without touching the borehole wall, compared to the rectangular fin, which has a wide tip at the end of the fin.

The primary aim of this study is to evaluate the thermal performance of the innovative U-Tube pipe configuration and compare it with the conventional single U-Tube type pipe configuration. In addition, the ground thermal properties for these two pipe configurations were estimated by using line source theory. In order to conduct TRTs, a small-scale experimental setup was designed, constructed, and commissioned in the laboratory at the University of Victoria, British Columbia, Canada.

2. Small-Scale Experimental Apparatus and Theory

2.1. Small-Scale Experimental Apparatus

The small-scale experimental apparatus comprises of three main parts: (i) VGHE (sand tank), (ii) water supply system, and (iii) data acquisition system.

A polyvinylchloride (PVC) borehole pipe (length, 1.05 m; inner diameter, 7.5 cm; outer diameter, 8.2 cm) is placed at the center from the top of a sand tank (length, 1 m; depth, 1.1 m; width, 1 m; wall thickness 2 cm), and the space between the borehole and the tank is filled with dry silica sand up to the top of the borehole. Researchers proposed that testing times can be significantly reduced by using a radius of influence in the range of 0.5 m to 1.5 m; this provides satisfactory temperature readings to determine thermal properties [26,27]. Figure 1 shows a photo of the small-scale test apparatus and Figure 2 is the schematic cross-section diagram of the same apparatus. Two different pipe configurations, including a conventional single U-Tube and an innovative U-Tube pipe configuration, are also inserted into the borehole at different times, and the space between the U-Tube pipe and the borehole wall is filled with grout (bentonite). Two thermistors (Omega TH-44032-1/4NPT-80; tolerance at 0–75 °C: ±0.1 °C) are installed at the inlet and outlet of the VGHE to measure the inlet and outlet fluid temperatures (see Figure 2). The connecting pipes between the sand tank and the bath are insulated to minimize the heat transfer between the connecting pipes and the ambient air.

The conventional single U-Tube pipe consists of two copper pipes that have an inner diameter of 12.7 mm and outer diameter of 16 mm with a length of 1 m. The innovative U-Tube pipe configuration consists of two copper pipes that have an inner diameter of 12.7 mm and outer diameter of 16 mm with a length of 1 m. The fins are made of copper and have a cross-section of 8 mm × 4 mm, attached over the full height (1 m) of the U-Tube. Figure 3 shows the cross-section of the two different pipe configurations. The parameters for the two different pipe configurations of VGHEs are listed in Table 1.

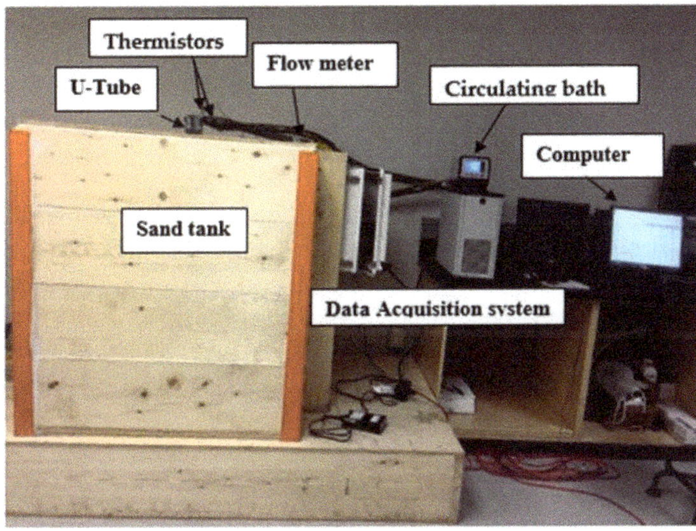

Figure 1. Photo of the small-scale experimental apparatus.

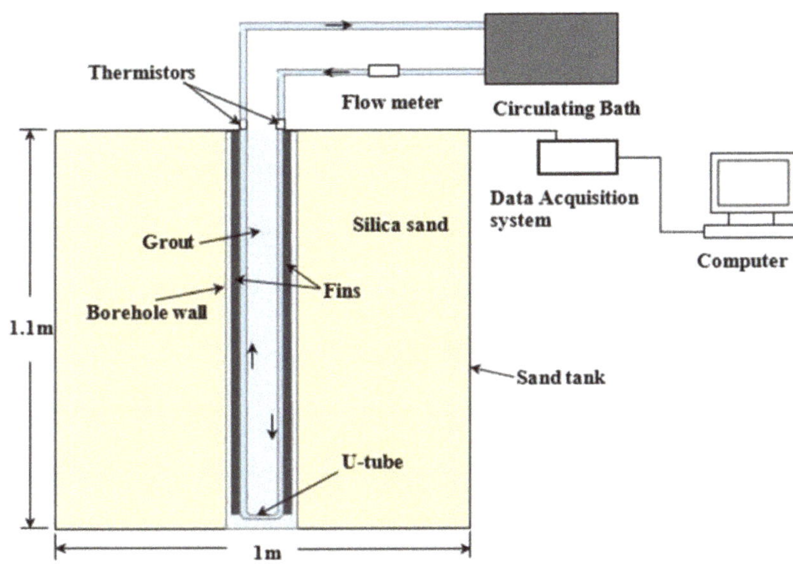

Figure 2. Schematic cross-section diagram of small-scale experimental apparatus.

A total of 64 thermocouples (EXPP-K-24) are distributed at specific distances inside the sand tank to measure the temperatures at different locations throughout the test and during the recovery time until the sand temperature reached the initial temperature. The thermocouples can measure a maximum temperature of 105 °C with an accuracy of ±0.75 °C. Each direction has 16 thermocouples, as shown in Figure 4.

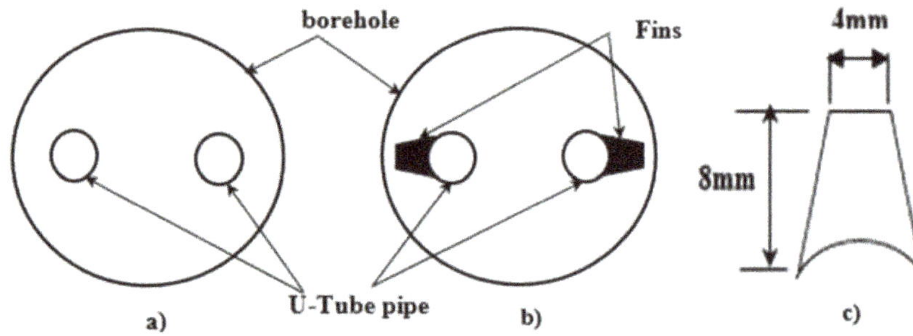

Figure 3. Cross-section of the two different pipe configurations and the fin: (**a**) conventional single U-Tube; (**b**) innovative U-Tube pipe configuration; (**c**) cross section of the fin.

Table 1. Parameters for the two different pipe configurations of VGHEs.

Parameters	Value	Unit
U-Tube pipe outer diameter	16	mm
U-Tube pipe inner diameter	12.7	mm
Space between U-Tube pipes	29	mm
Length of the U-Tube pipe	1.05	m
Thermal conductivity of copper pipe [28]	400	W/m·K
Fin width	4	mm
Fin length	8	mm
Thermal conductivity of copper fin [28]	400	
Borehole outer diameter (PVC)	82	mm
Borehole inner diameter (PVC)	75	mm
Thermal conductivity borehole wall (PVC) [28]	0.19	W/m·K
Thermal conductivity bentonite (grout) [29]	0.8	W/m·K
Thermal conductivity of silica sand (ground/soil) [12]	2.42	W/m·K
Specific heat of silica sand [12]	750	J/kg. K
Density of silica sand [12]	1700	Kg/m^3

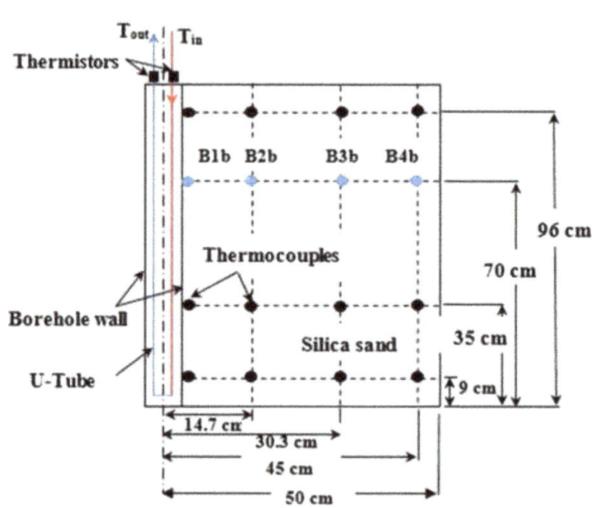

Figure 4. Front view for the 16 thermocouples at borehole wall and in silica sand at specific locations on the right side of sand tank.

2.2. Theory

The goal of conducting TRT is to estimate the thermal (i.e., heat transfer) efficiency of the VGHE and also the thermal properties of the ground in field applications. The rate of the heat transfers per unit length (depth) of VGHE can be calculated from the relationship between the measured inlet and the outlet temperatures of the heat carrier fluid (water), which circulates through the VGHE, and its flow rate, as shown in Equation (1) [12,24,30,31]:

$$q = \frac{C_p\ \dot{m}\ (T_{in} - T_{out})}{H} \tag{1}$$

where T_{in} is the inlet fluid temperature (°C), T_{out} is the outlet fluid temperature (°C), q is the heat exchange rate per unit borehole length (W/m), C_p is the specific heat capacity of the circulating fluid (J/kg·K), $\dot{m}$ mass flow rate (kg/s), and H is the borehole depth (m).

To estimate the ground thermal properties, the line source model (LSM) is used in this research. Ingersoll and Plass (1948) reported that the infinite line source was used in the 1940s to calculate the ground temperature change over time for ground loop heat exchangers. The first field study was carried out to estimate the thermal conductivity of the ground by using this theory [32].

The temperature change in the ground is a function of radius (r) from the borehole center and time (t) with an injection of a constant amount of heat (q) from the top to the bottom of the borehole that can be calculated by using the following equation [33–36]:

$$T(r,t) = \frac{Q}{4\pi\lambda_{eff}H}\int_{\frac{r}{2\pi k}}^{\infty}\frac{e^{-\beta^2}}{\beta}d\beta = \frac{Q}{4\pi\lambda_{eff}H}E_1\left(\frac{r^2}{4\alpha t}\right) \tag{2}$$

where λ_{eff} is the effective ground thermal conductivity (W/m·K), and α is the thermal diffusivity of the ground m^2/s. When the values of the parameter $\frac{\alpha t}{r^2}$ are large, it can be expressed with the following simple relation:

$$E_1\left(\frac{\alpha t}{r^2}\right) = ln\left(\frac{4\alpha t}{r^2}\right) - \gamma \tag{3}$$

where, E_1 is the exponential integral and γ is Euler's constant (0.57721).

It is reported that the maximum error resulting from using the value of $\frac{\alpha t}{r^2} \geq 5$ is 10%, and it is reduced to 2.5% when the value of $\frac{\alpha t}{r^2} \geq 20$ is used [35,36].

The line source temperature at the borehole radius (r_b), including the impact of the borehole thermal resistance (R_b) between the heat carrier fluid and the borehole wall, was used to evaluate the mean fluid temperature $(T_f(t))$. The fluid temperature as a function of time could be written in Equation (4) [30,33,35]:

$$T_f(t) = \frac{Q}{4\pi\lambda_{eff}H}ln(t) + \frac{Q}{H}\left[\frac{1}{4\pi\lambda_{eff}}\left\{ln\left(\frac{4\alpha}{r_b^2}\right) - \gamma\right\} + R_b\right] + T_o \tag{4}$$

where R_b is the borehole thermal resistance (m·K/W) and T_o is the initial ground temperature (°C).

The borehole thermal resistance can be expressed by the following equation:

$$R_b = \frac{H}{Q}\left[T_f(t) - T_o\right] - \frac{1}{4\pi\lambda_{eff}}\left[\ln(t) + ln\left(\frac{4\alpha}{r_b^2}\right) - \gamma\right] \tag{5}$$

It is noticed that the second and third terms on the right side of Equation (4) are constant. Therefore, it can be considered that the above equation is similar to the linear slope equation as $y = m.x + b$.

It is assumed that there is a linear relationship between the mean fluid temperature and the logarithm time $\ln(t)$ [33,36,37]. They also expressed this relation by using the following equation:

$$T_f(t) = m.ln(t) + b \quad (6)$$

$$Slope = m = \frac{q}{4\pi\lambda_{eff}} \rightarrow \lambda_{eff} = \frac{q}{4\pi m} \quad (7)$$

3. Analysis of Experimental Results

Effective Ground Thermal Conductivity and Borehole Thermal Resistance

Two different approaches can be used to estimate the initial ground temperature before the TRT started [38]. The first approach is to calculate the initial ground temperature with still water in the U-Tube. The inlet and outlet borehole temperatures and the ground temperature at 64 locations in the sand tank are recorded every 1 min for about 1 h by using the data acquisition system. The second approach is to circulate the water in a close system through the VGHE for about 1 h and without adding heat during the test. Nevertheless, there was no heat injection during this period, but at the same time, some heat would be added into the system by the pump. Then, the inlet and outlet borehole temperatures and the ground temperatures at the 64 locations in the sand tank were recorded every 1 min for 1 h by using the data acquisition system. In this study, the first approach is used to estimate the average initial ground temperature. The U-Tube in the VGHE wall was filled with water for 7 days before the measurement started. The average initial ground temperature was calculated to be 17.58 °C.

After the average initial ground temperature was calculated as shown before by using first approach, the TRT was conducted for 130 h (60 h for the heat injection and 70 h to let the sand return to its initial ground temperature). A constant heat injection rate was used throughout the TRT. The outlet fluid temperature leaving from the circulating bath and entering to the VGHE was set to be 50 °C with a constant volumetric flow rate (V) of 0.730 L/min. The inlet and outlet fluid temperatures of the borehole were recorded every 5 min by using the data acquisition system. As shown in Figure 5, the difference between the inlet and outlet temperatures was 0.82 °C after 1h. The temperature difference then gradually decreased to 0.52 °C after 30 h, and it decreased further to 0.4 °C after 60 h.

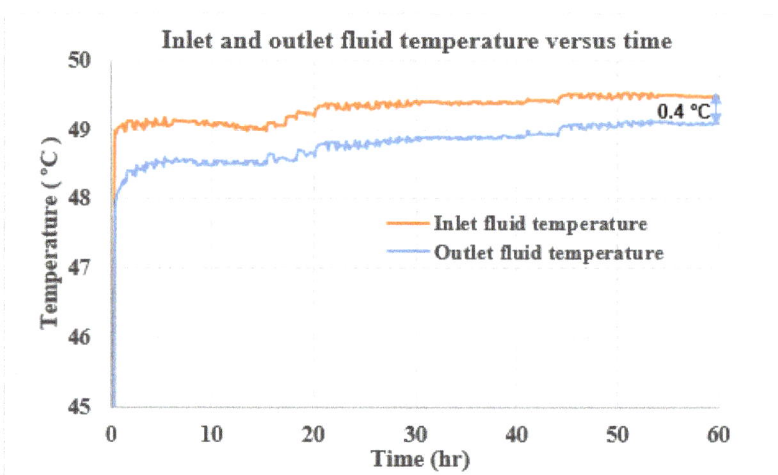

Figure 5. Inlet and outlet fluid temperature for the conventional single U-Tube VGHE during heat injection.

As shown in Equation (6), there is a linear relationship between T_f and $ln(t)$. The slope (m) = 0.6189 was calculated from Figure 6, and then substituting 'm' in the Equation (7), λ_{eff} was calculated as 3.92 W/m·K.

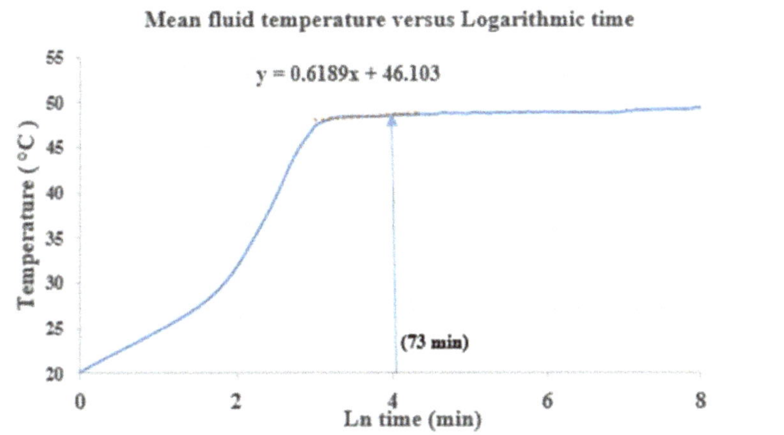

Figure 6. Mean fluid temperature plotted versus logarithmic time, conventional single U-Tube VGHE.

Using Equation (1) and from the measured inlet (T_{in}) and outlet (T_{out}) temperatures, the heat exchange rate per unit length (q) of VGHE was calculated. The borehole thermal resistance (R_b) was calculated using the value of λ_{eff} (3.92 W/m·K) in Equation (5). The value of borehole thermal resistance (R_b) at $t > 5\ r2/\alpha$ (i.e., 73 min, error 10%), was found to be 0.961 m·K/W. Calculated results are presented in Table 2.

Table 2. Comparison values of ΔT, V, q, λeff, and R_b for the two pipe configurations.

Pipe Configurations	ΔT at 60 h °C	V (L/min)	q at 60 h (W/m)	λ*eff* (W/m·K)	R_b (m·K/W)
Single U-Tube pipe	0.4	0.729	19.05	3.92	0.961
innovative single U-Tube	0.7	0.703	30.09	4.85	0.680

Four thermocouples, B1b, B2b, B3b, and B4b, were fixed on the right hand side of the borehole wall at 14.70, 30.30, and 45.00 cm, respectively, from the center of the borehole, at a height of 700 cm from the bottom of the sand tank. The four thermocouples (EXPP-K-24) measure the maximum temperature of 105 °C with an accuracy of ±0.75 °C. Before conducting the test, the initial ground temperature of B4b, B3b, B2b, and B1b were 17.75 °C, 17.98 °C, 18.06 °C, and 18.09 °C, respectively, and they increased after 60 h during the heat injection to 36.16 °C, 25.73 °C, 20.52 °C, and 19.95 °C, respectively. The ground temperatures of the B4b, B3b, B2b, and B1b decreased to 18.2 °C, 18.28 °C, 18, and 17.71 °C, respectively, after 70 h from the end of the test, as shown in Figure 7. However, the ground temperature changed rapidly and reached 36.16 °C after 60 h at the borehole wall. It changed slowly away from the borehole at 45 cm from the center of the borehole and reached 19.95 °C. The ambient temperature was recorded during the TRT and recovery time by using a dada logger (OM 62), and it was confined at a range of 14.64 °C to 20.48 °C. The temperature measurement for the OM-62 was at a range of −40 °C to 70 °C with an accuracy of ±0.5.

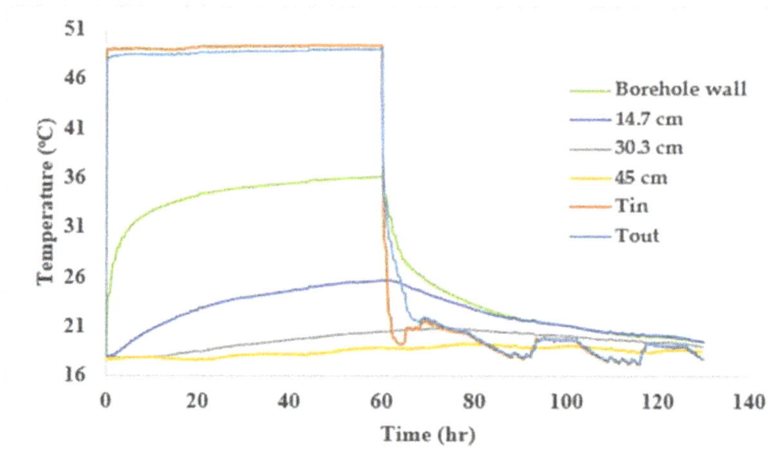

Figure 7. Temperature measurement at the inlet (Tin) and the outlet (Tout), borehole wall, and in sand at different locations from the borehole wall, conventional single U-Tube VGHE.

Another TRT was performed for the innovative U-Tube pipe configuration with the similar boundary conditions used with the conventional single U-Tube. The average initial ground temperature was calculated to be 19.91 °C before the TRT started.

The TRT was conducted for 145 h (65 h for the heat injection and 80 h to let the system return to its initial ground temperature). A constant heat injection rate was used throughout the TRT. The outlet fluid temperature leaving from the bath and entering to the VGHE was set to be 50 °C with a constant volumetric flow rate of 0.703 L/min. The difference between the inlet and outlet temperatures was 0.97 °C after1 h. The temperature difference then gradually decreased to 0.7 °C after 30 h, and it got to 0.65 °C after 60 h as shown in Figure 8.

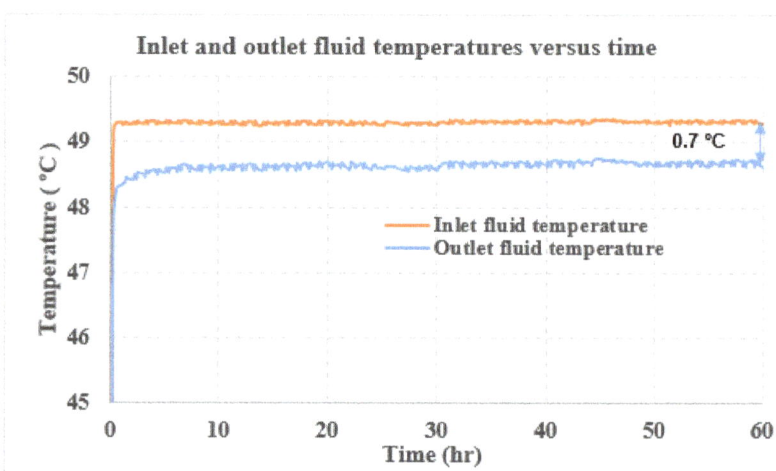

Figure 8. Inlet and outlet fluid temperature for the innovative U-Tube pipe VGHE, during heat injection.

As shown in Equation (6), there is a linear relationship between T_f and $ln(t)$. The slope (m) = 0.6642 was calculated from Figure 9, and then substituting 'm' in the Equation (7), $\left(\lambda_{eff}\right)$ was calculated as 4.85 W/m·K.

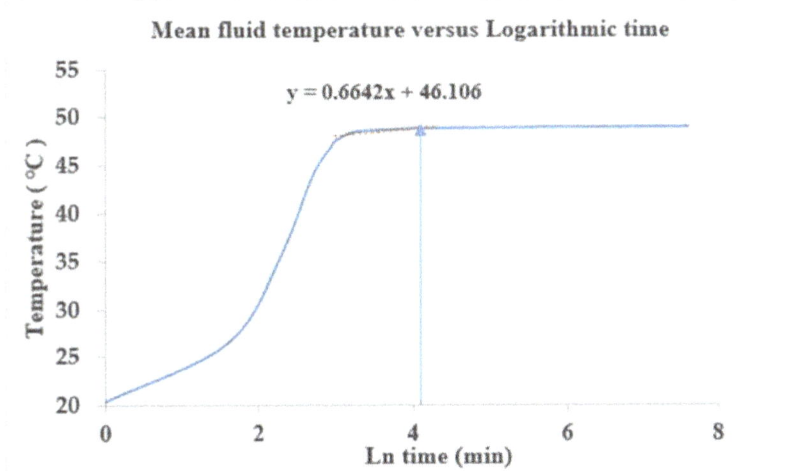

Figure 9. Mean fluid temperature plotted versus logarithmic time, innovative U-Tube pipe VGHE.

Using Equation (1) and from the measured inlet (Tin) and outlet (Tout) temperatures, the heat exchange rate per unit length (q) of VGHE was calculated. The value of borehole thermal resistance (R_b) was calculated using the value of λ_{eff} (4.85 W/m·K) in Equation (5). The value of borehole thermal resistance (R_b) at t > 5 r2/α (i.e., 73 min, error 10%), was found to be 0.680 m·K/W. Calculated results are presented in Table 2.

As shown in Figure 10, the initial ground temperature of B4b, B3b, B2b, and B1b were 21.08 °C, 22.02 °C, 21.72 °C, and 21.09 °C, respectively, and they increased after 60 h during the heat injection to 41.03 °C, 29.28 °C, 23.71 °C, and 21.69 °C, respectively. The ground temperatures of the B4b, B3b, B2b, and B1b decreased to 20.26 °C, 20.55 °C, 19.74 °C, and 18.5 °C, respectively, after 80 h from the end of the test as shown in Figure 10. However, the ground temperature changed rapidly and reached to 41.03 °C after 60 h at the borehole wall. It changed slowly away from the borehole at 45 cm from the center of the borehole and reached 21.69 °C. The ambient temperature was recorded during the TRT and recovery time by using a data logger (OM-62), and it was confined at a range of 14.84 °C to 21.32 °C.

Table 2 presents the difference between the inlet and outlet fluid temperatures (ΔT), volumetric flow rate (V), heat injection rate (q), effective ground thermal conductivity (λ_{eff}), and borehole thermal resistance (R_b) for the two different pipe configurations.

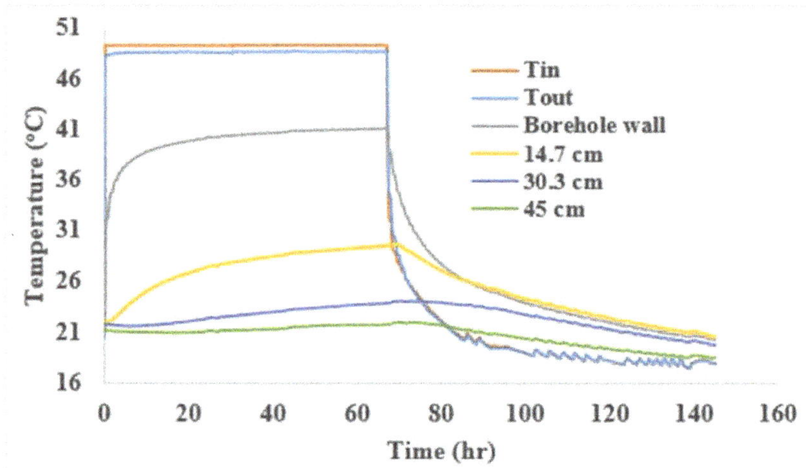

Figure 10. Temperature at the inlet (Tin) and the outlet (Tout), borehole wall, and in sand at different locations from the borehole wall, innovative U-Tube pipe VGHE.

4. Discussion

4.1. Experimental Observations

Two TRTs were conducted in the laboratory for the two different pipe configurations under similar boundary conditions to demonstrate the impact of innovative U-Tube pipe configuration on the heat transfer rate of the VGHE. The two different pipe configurations are the conventional single U-Tube and the innovative pipe configuration (the conventional single U-Tube pipe with fins), as shown in Figure 3. The two TRTs are continued for a period of 60 h. As shown in Figures 5 and 8, after 60 h from starting the test, the difference between the inlet and outlet fluid temperatures was 0.4 °C for the conventional single U-Tube pipe VGHE while it was 0.7 °C for innovative U-Tube pipe configuration VGHE. The latter was increased by 0.29 °C. As would be expected, the borehole thermal resistance also decreased from 0.961 m·K/W to 0.680 m·K/W due to the introduction of external fins to the conventional single U-Tube (see Table 2). The effective ground thermal conductivity is 3.92 W/m·K for the conventional single U-Tube pipe configuration, and it is 4.85 W/m·K for the innovative U-Tube pipe configuration. The results also indicate that that the heat exchange rate for the innovative U-Tube pipe configuration was 30.09 W/m, i.e., an increase of 11.04 W/m compared to the conventional single U-Tube (see Table 2). The results also show that the temperature at the borehole wall increased by 4.87 °C when the innovative U-Tube pipe configuration was used compared to the conventional single U-Tube. With the innovative U-Tube pipe configuration, at 45 cm from the center of the borehole and at different depths, the highest temperature increased by 0.54 °C more than the conventional single U-Tube pipe configuration as shown in Figures 7 and 10. Based on the aforementioned observations, it can be concluded that substantial improvement of thermal efficiency of VGHE can be made by the introduction of external fins to the conventional single U-Tube pipe. In practical terms, it means fewer boreholes are required when conventional single U-Tube pipes are replaced by the innovative U-Tube pipe configuration with external fins, and this, in turn, will reduce the VGHE installation cost.

4.2. Sources of Uncertainty or Error

The sources of error or uncertainty in measured parameters during the thermal response tests (TRTs) and their impacts on the results obtained from the tests in the laboratory could be explained qualitatively but difficult to quantify adequately. These sources of error arise from the assumptions made regarding the properties of the materials, application of the line source theory, sensors, etc. [39]. Quantification of uncertainties arising from these sources is not only beyond the scope of this paper, but a systemic evaluation protocol for this purpose is yet to be established. However, it is well known that the errors associated with measured inlet and outlet temperatures, using two thermistors (Omega TH-44032-1/4NPT-80), have a significant impact on the overall error or uncertainty of the test results [39]. The error associated with each of these two measured temperatures is ±0.1 °C at 0 to 75 °C. The propagation of this error due to subtraction can be calculated by square rooting the sum of the square of each errors (i.e., $\sqrt{(0.1)^2 + (0.1^2)} = \pm 0.14$ °C).

5. Conclusions

The results from the TRTs conducted in the laboratory on a small-scale VGHE with two different pipe configurations, a conventional single U-Tube and the same with external fins, have been presented and examined in this paper. The main observations are outlined below:

- The difference between the inlet and outlet temperature for the innovative U-Tube pipe configuration (i.e., with external fins) is 0.7 °C after 60 h while the difference between the inlet and outlet temperatures for the conventional single U-Tube pipe configuration was 0.4 °C after 60 h. Thus, a 76.3% increase in temperature difference has been achieved due to the use of innovative pipe configuration.
- The effective ground thermal conductivity for the innovative U-Tube pipe configuration was 4.85 W/m·K, a 23.64% increase compared to the conventional single U-Tube pipe configuration.
- The borehole thermal resistance for the innovative U-Tube pipe configuration was found to be 0.680 m·K/W, which is 29.22% lower than that of the same with conventional single U-Tube pipe.
- The increased ground temperatures with the innovative U-Tube pipe configuration, compared to single U-Tube pipe configuration, clearly indicate the superior heat transfer performance of innovative pipe configuration.
- The introduction of external fins to single U-Tube pipe configuration helps to reduce the number of boreholes by about 57.95%, which will in turn decrease the installation cost.

Author Contributions: Conceptualization, A.E. and P.M.; methodology, A.E.; formal analysis, A.E. and P.M.; investigation, A.E.; resources, P.M.; writing—original draft preparation, A.E. and P.M.; writing—review and editing, P.M.; supervision, P.M.; project administration, P.M.; funding acquisition, P.M. All authors have read and agreed to the published version of the manuscript.

Funding: The graduate student (A.E.) was funded by the Libyan-North American Scholarship Program (LNASP), which is a joint collaboration between the Libyan Ministry of Higher Education and Scientific Research and the Canadian Bureau for International Education.

Institutional Review Board Statement: Not applicable.

Informed Consent Statement: Not applicable.

Data Availability Statement: Not applicable.

Acknowledgments: The authors would like to acknowledge the scientific and technical knowledge supports provided by Andrew Rowe, Armando Tura, Mitchell Anderson, Matthew Walker, and Robert Lepage.

Conflicts of Interest: The authors declare no conflict of interest. The funders had no role in the design of the study; in the collection, analyses, or interpretation of data; in the writing of the manuscript, or in the decision to publish the results.

References

1. Muraya, K.N. Numerical Modeling of the Transient Thermal Interference of Vertical U-Tube Heat Exchangers. Ph.D. Thesis, Texas A&M University, College Station, TX, USA, 1994.
2. Bidarmaghz, A.; Narsilio, G.; Johnston, I. Numerical modelling of ground heat exchangers with different ground loop configurations for direct geothermal applications. In Proceedings of the 18th International Conference on Soil Mechanics and Geotechnical Engineering 2013, Paris, France, 2–6 September 2013; pp. 2–6.
3. Sanner, B.; Hellström, G.; Spitler, J.; Gehlin, S. Thermal response test–current status and world-wide application. In Proceedings of the World Geothermal Congress, Antalya, Turkey, 24–29 April 2005; Volume 1436, p. 2005.
4. Gu, Y.; O'Neal, D.L. Modeling the effect of backfills on U-tube ground coil performance. *Ashrae Trans.* **1998**, *104*, 356.
5. Pahud, D.; Matthey, B. Comparison of the thermal performance of double U-pipe borehole heat exchangers measured in situ. *Energy Build.* **2001**, *33*, 503–507. [CrossRef]
6. Zeng, H.; Diao, N.; Fang, Z. Heat transfer analysis of boreholes in vertical ground heat exchangers. *Int. J. Heat Mass Transf.* **2003**, *46*, 4467–4481. [CrossRef]
7. Esen, H.; Inalli, M. Temperature distributions in boreholes of a vertical ground-coupled heat pump system. *Renew. Energy* **2009**, *34*, 2672–2679. [CrossRef]
8. Acuña, J.; Palm, B. A novel coaxial borehole heat exchanger: Description and first distributed thermal response test measurements. In Proceedings of the World Geothermal Congress 2010, Bali, Indonesia, 25–29 April 2010; p. 7.
9. Lee, C.; Park, M.; Min, S.; Kang, S.-H.; Sohn, B.; Choi, H. Comparison of effective thermal conductivity in closed-loop vertical ground heat exchangers. *Appl. Therm. Eng.* **2011**, *31*, 3669–3676. [CrossRef]
10. Beier, R.A.; Ewbank, G.N. *In-Situ Test Thermal Response Tests Interpretations. OG&E Ground Source Heat Exchange Study*; Oklahoma State University: Stillwater, OK, USA, 2012.
11. Desmedt, J.; Van Bael, J.; Hoes, H.; Robeyn, N. Experimental performance of borehole heat exchangers and grouting materials for ground source heat pumps. *Int. J. Energy Res.* **2012**, *36*, 1238–1246. [CrossRef]
12. Haddada, J.; Miyara, A. Performance Investigation of Multiple-Tube Ground Heat Exchangers for Ground-Source Heat Pump. *Am. J. Energy Eng.* **2014**, *2*, 103–107.
13. Kramer, C.A.; Basu, P. Performance of a model geothermal pile in sand. In Proceedings of the 8th International Conference on Physical Modelling in Geotechnics 2014 (ICPMG2014), Perth, Australia, 14–17 January 2014; Gaudin, C., White, D., Eds.; CRC Press/Balkema: Leiden, The Netherlands, 2014; pp. 771–777.
14. Salim Shirazi, A.; Bernier, M. A small-scale experimental apparatus to study heat transfer in the vicinity of geothermal boreholes. *HVAC&R Res.* **2014**, *20*, 819–827.
15. Erol, S.; François, B. Efficiency of various grouting materials for borehole heat exchangers. *Appl. Therm. Eng.* **2014**, *70*, 788–799. [CrossRef]
16. Cimmino, M.; Bernier, M. Experimental determination of the g-functions of a small-scale geothermal borehole. *Geothermics* **2015**, *56*, 60–71. [CrossRef]
17. Liu, X.; Xiao, Y.; Inthavong, K.; Tu, J. Experimental and numerical investigation on a new type of heat exchanger in ground source heat pump system. *Energy Effic.* **2015**, *8*, 845–857. [CrossRef]
18. Ramadan, A.S. Parametric Study of Vertical Ground Loop Heat Exchangers for Ground Source Heat Pump Systems. Master's Thesis, University of Western Ontario, London, ON, Canada, 2016.
19. Chang, K.S.; Kim, M.J. Thermal performance evaluation of vertical U-loop ground heat exchanger using in-situ thermal response test. *Renew. Energy* **2016**, *87*, 585–591. [CrossRef]
20. Luo, J.; Zhao, H.; Gui, S.; Xiang, W.; Rohn, J.; Blum, P. Thermo-economic analysis of four different types of ground heat exchangers in energy piles. *Appl. Therm. Eng.* **2016**, *108*, 11–19. [CrossRef]
21. Serageldin, A.A.; Sakata, Y.; Katsura, T.; Nagano, K. Thermo-hydraulic performance of the U-tube borehole heat exchanger with a novel oval cross-section: Numerical approach. *Energy Convers. Manag.* **2018**, *177*, 406–415. [CrossRef]
22. Bae, S.M.; Nam, Y.; Choi, J.M.; Lee, K.H.; Choi, J.S. Analysis on thermal performance of ground heat exchanger according to design type based on thermal response test. *Energies* **2019**, *12*, 651. [CrossRef]
23. Li, W.; Li, X.; Peng, Y.; Wang, Y.; Tu, J. Experimental and numerical investigations on heat transfer in stratified subsurface materials. *Appl. Therm. Eng.* **2018**, *135*, 228–237. [CrossRef]
24. Liang, B.; Chen, M.; Fu, B.; Li, H. Investigation on the thermal and flow performances of a vertical spiral-tube ground heat exchanger in sand combined with kaolin additive. *Energy Build.* **2019**, *190*, 235–245. [CrossRef]
25. Eswiasi, A. Novel Pipe Configuration for Enhanced Efficiency of Vertical Ground Heat Exchanger. Ph.D. Thesis, University of Victoria, Victoria, BC, Canada, 2021.
26. Raymond, J.; Therrien, R.; Gosselin, L.; Lefebvre, R. A Review of Thermal Response Test Analysis Using Pumping Test Concepts. *Ground Water* **2011**, *49*, 932–945. [CrossRef] [PubMed]
27. Zhang, C.; Guo, Z.; Liu, Y.; Cong, X.; Peng, D. A review on thermal response test of ground-coupled heat pump systems. *Renew. Sustain. Energy Rev.* **2014**, *40*, 851–867. [CrossRef]
28. Salimshirazi, A. Transient Heat Transfer in Vertical Ground Heat Exchangers. Ph.D. Thesis, École Polytechnique de Montréal, Montreal, QC, Canada, 2012.
29. Geothermal Grout™, Technical Data, Cetco Building Products. Available online: www.cetco.com/dpg (accessed on 3 June 2021).

30. Choi, W.; Ooka, R. Effect of natural convection on thermal response test conducted in saturated porous formation: Comparison of gravel-backfilled and cement-grouted borehole heat exchangers. *Renew. Energy* **2016**, *96*, 891–903. [CrossRef]
31. Zhou, T.; Chen, M.; Liang, B. Thermal performance of a ground U-shaped tube with twisted tapes in sand/graphite backfill materials. *Exp. Heat Transf.* **2021**, *34*, 186–200. [CrossRef]
32. Mogensen, P. Fluid to duct wall heat transfer in duct system heat storages. *Doc. Swed. Counc. Build. Res.* **1983**, 652–657.
33. Eklöf, C.; Gehlin, S. *TED-a Mobile Equipment for Thermal Response Test: Testing and Evaluation*; Luleå University of Technology: Luleå, Sweden, 1996. Available online: https://www.researchgate.net/publication/242072965_A_Mobile_Equipment_for_Thermal_Response_Test (accessed on 3 June 2021).
34. Austin, W.A., III. Development of An In Situ System for Measuring Ground Thermal Properties. Ph.D. Thesis, Oklahoma State University, Stillwater, OK, USA, 1998.
35. Gehlin, S. Thermal Response Test: Method Development and Evaluation. Ph.D. Thesis, Department of Environmental Division of Water Resources Engineering, Lulea University, Lulea, Sweden, 2002.
36. You, J.; Lee, C. Analysis of effective thermal conductivity stability of geothermal heat exchanger according to ambient environmental conditions. *Int. J. Energy Res.* **2019**, *43*, 7682–7692. [CrossRef]
37. Loveridge, F.; Olgun, C.G.; Brettmann, T.; Powrie, W. The Thermal Behaviour of Three Different Auger Pressure Grouted Piles Used as Heat Exchangers. *Geotech. Geol. Eng.* **2015**, *33*, 273–289. [CrossRef]
38. Gehlin, S.; Nordell, B. Determining undisturbed ground temperature for thermal response test. *ASHRAE Trans.* **2003**, *109*, 151–156.
39. Witte, H.J.L. Error analysis of thermal response tests. *Appl. Energy* **2013**, *109*, 302–311. [CrossRef]

sustainability

MDPI

Article

Lookup Tables for Power Generation Performance of Photovoltaic Systems Covering 40 Geographic Locations (Wilayats) in the Sultanate of Oman, with and without Solar Tracking, and General Perspectives about Solar Irradiation

Osama A. Marzouk

College of Engineering, University of Buraimi, Al Buraimi 512, Oman; osama.m@uob.edu.om

Abstract: An energy modeler for solar photovoltaic (PV) systems may be limited to climatic data of certain major cities, not covering the one for which the PV system is intended. Additionally, a person not skilled in solar PV modeling may still desire a quick estimate of PV system electricity generation to help decide the level of investment in PV systems. This work addresses these points by establishing lookup tables to summarize predicted electricity generation, solar irradiation, and optimum orientation at various locations in the Sultanate of Oman. The results are produced by processing simulation data using the online open-access tool PVGIS (Photovoltaic Geographical Information System) of the European Commission's Joint Research Centre (EC-JRC). The tables cover 40 out of the country's 61 s-level administrative divisions (wilayats) and cover fixed and movable PV panels. The results show that the yearly electricity generation can change up to 11.86% due to the change of location. Two-axis PV tracking offers a small improvement (about 4% on average) over single-vertical-axis tracking but offers noticeable improvement (about 34% on average) over optimally oriented fixed PV panels. Monthly profiles of expected PV electricity generation, as well as the generation drop due to changing the PV mounting from free standing to building integrated, were examined for three locations. As general perspectives that may be of interest to global readers, this work provides quantitative evidence of the overall accuracy of the PVGIS-SARAH database through comparison with ground-measured global horizontal irradiation (GHI). In addition, a full example is presented considering 12 different countries in the northern and southern hemispheres that brings the attention of solar energy modelers to the level of errors they may encounter when the impact of longitude (thus, the exact location) is ignored for simplicity, while focus is given to the latitude.

Keywords: lookup tables; photovoltaic; PVGIS; solar; Oman

Citation: Marzouk, O.A. Lookup Tables for Power Generation Performance of Photovoltaic Systems Covering 40 Geographic Locations (Wilayats) in the Sultanate of Oman, with and without Solar Tracking, and General Perspectives about Solar Irradiation. *Sustainability* **2021**, *13*, 13209. https://doi.org/10.3390/su132313209

Academic Editor: Mostafa Ghasemi Baboli

Received: 29 September 2021
Accepted: 17 November 2021
Published: 29 November 2021

1. Introduction

Limiting the global temperature increase to 2 °C above pre-industrial levels (before the intensive industrial activities that started around 1750) demands reaching net-zero carbon emissions by around 2070, while limiting this to 1.5 °C demands an earlier attainment of the net-zero emissions of carbon dioxide by around 2050 [1]. Decarbonization efforts include integrating renewable energy electricity generation from solar photovoltaic (PV) panels, thereby reducing electricity generation from fossil fuels that lead to carbon emissions [2]. In 2019, solar photovoltaic technology contributed about 679 TWh to electricity generation, forming a share of 9.8% among renewable energy sources [3]. The transforming energy scenario of IRENA (International Renewable Energy Agency) calls for an increased share of 25% for the solar photovoltaic technology, representing 13,787 TWh in 2050, in order to restrain the global temperature increase to 1.5 °C and bring carbon dioxide emissions close to net zero by 2050 [4]. It is appealing to see that solar PV power generation increased by 22% in 2019, taking the second position in terms of generation growth among renewable energy technologies (with wind energy technology being in the first position, leading solar PV energy technology by a small gap). Extrapolating the nonlinear accelerating progress in solar PV technology over the period of 2006–2020 suggests that it meets the Sustainable Development Scenario (SDS) of the IEA (International Energy Agency), reaching 3269 TWh in 2030, compared to only 6 TWh in 2006 [5]. Such intensive deployment of solar PV technology as well as the demand for continuous expansion in its utilization warrants a parallel effort in numerically simulating PV systems ahead of installation so that a PV system can be reasonably sized according to the demanded electricity generation at the specific geographic location of operation.

Mathematical models that describe the solar radiation at an arbitrary time or location exist [6–8], as well as software programs that enable prediction of the generated electricity from a PV design [9–11]. When predicting the electricity generation from complex systems (as compared to a single PV panel or a solar cell) under optimized configuration throughout the year, mathematical equations become impractical alone, and one needs to resort to a computational tool that accounts for different factors affecting the performance of the PV system, while yielding a result in a reasonably short time. Even with the availability of software programs for simulating PV systems, a climatic record is needed for useful prediction. The climatic information is location-dependent, and the simulation tool may come with a database of locations for the user to select from as a convenient way of providing the necessary meteorological data. In the case where the users are interested in a location not in the built-in database, they may need to set the environment to the nearest location available, or the simulation tool may perform interpolation. Either option incurs some approximation that can introduce errors. It can be helpful if pre-calculated one-time lookup tables are made for users, with multiplication factors that help in translating the predicted electricity from a PV system of a certain design type (such as fixed PV panels, single-axis tracking PV panels, and two-axis tracking PV panels) that have records for a large number of geographic locations or communities within a country. Then, a PV system modeler who already predicted the performance of a PV system in one location may use the table to translate the predicted electricity at different location by using the proper factors in the lookup tables that correspond to the specific type of concern. This procedure extends the capability of the simulation beyond the embedded database of locations it has. The lookup tables can be established for arbitrary locations with a spatial resolution that can easily be expanded when needed. They are also not tied to a specific PV simulation tool, simply being tables that correct the yearly PV electricity generation at one location (where the electricity generation is known) so that it corresponds to another location (where electricity generation is sought). In addition, personnel interested in PV systems may use the tables without the need of performing any solar simulation. This relieves the need to have access to a PV simulation tool or to have the technical skills to use it, which may not be readily available to non-experienced persons. The known yearly electricity generation (to be projected or corrected for another location) does not have to come from a simulation tool; instead, it can be an actual measured value [12,13]. In this case, the lookup tables are still useful in providing a quantitative assessment of the performance of such a PV system if replicated in another location. Aside from conveniently projecting the yearly electricity generation to different geographic locations, the lookup tables themselves can include additional location-specific valuable information to PV investors in the public or private sectors (such as the solar irradiation, the expected electricity generation for a standard PV size, and the optimum orientation for fixed PV panels) and also enable convenient quantitative identification of locations of the highest potential for solar power generation, which can be given a priority over other locations. In relation to the previous remarks, this work gives three lookup tables for use in the Sultanate of Oman, corresponding to three categories of PV solar systems, which are:

- Fixed PV panels with optimum azimuth and tilt;
- Moving PV panels with vertical-axis solar tracking;
- Moving panels with two-axis solar tracking.

The Sultanate of Oman (or simply Oman) is administratively divided into 61 s-level administrative divisions ("wilayats" according to the local language) in 11 governorates. The country enjoys relatively high solar radiation [14], making solar energy a primary option for the country in case it decides to reshape its energy mix, having a significant share of renewable energy in it, including solar PV energy, consistent with the country-led renewables readiness assessment (RRA) and long-term electricity transmission system master plan, having an outlook toward 2070 [15]. A map of Oman is given in Figure 1, showing the 11 governorates. The capital Muscat lies in the governorate of Muscat, which is also the name of one of the second-level administrative divisions (wilayats) within that governorate. The figure also highlights the location of three wilayats, in addition to Muscat, which are: Thumrait (in the southern governorate of Dhofar), Madha (in the northern governorate of Musandam), and Adam (in the central governorate of Ad Dakhliya). This can facilitate later discussions about these specific locations, where the first one shows the best PV performance among the 40 analyzed wilayats, the second one shows the poorest performance among the 40 analyzed wilayats, and the third one shows third-party ground-based measurements of monthly global horizontal irradiation (GHI) that are used in validation. For readers not familiar with Oman, these additional details can be very helpful.

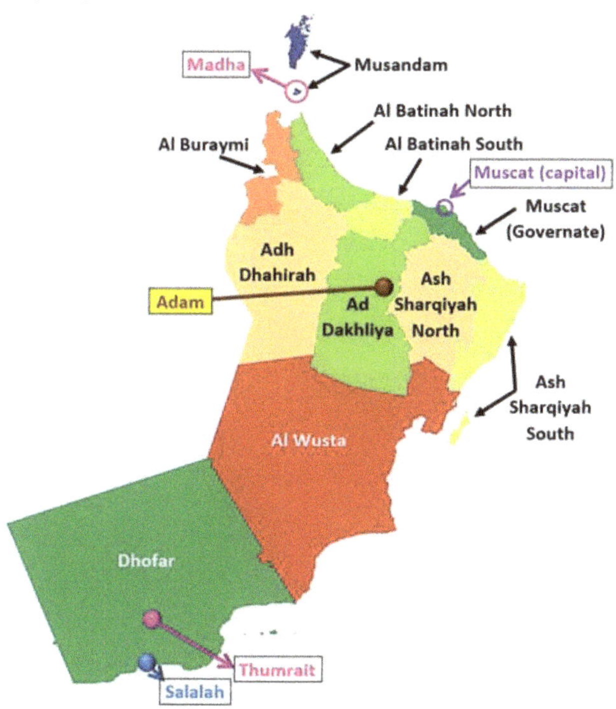

Figure 1. A map of the Sultanate of Oman, showing its 11 first-level administrative divisions (governorates) as well as its capital (Muscat). It also shows two locations which receive special attention in this work as having the highest and the lowest predicted PV performance among all 40 examined locations, irrespective of whether solar tracking (single-vertical-axis or two-axis) is enabled or not (Thumrait—highest performance, and Madha—lowest performance). In addition, the location of Adam is also highlighted, where comparison with ground-measurement irradiation data is made. The location of Salalah is indicated as well, whose horizon height profile is discussed in a subsequent part of this work. The map is adapted from two maps available freely and licensed under the Open Government License of the Sultanate of Oman [16,17].

The governorates of Oman vary not only in land size, but also in population, terrain, and climate. The governorate of Musandam is an exclave located in the north tip of Oman.

This work considers 40 out of the 61 s-level administrative divisions (the wilayats) of Oman. For simplicity, and since a single point was analyzed per second-level administrative division, they are referred to here as "locations", which, for the purpose of this work, serves to indicate the geographic points that represent a second-level administrative division selected for photovoltaic analysis and inclusion in each of the lookup tables. In some regions, the locations are relatively close to each other, so not all of them were included in the analysis. However, the number of locations selected for analysis in each governorate never drops below half the number of wilayats in that governorate. Considering 40 wilayats out of the total 61 wilayats appears to give good coverage for Oman. Expanding the analysis to the remaining 21 wilayats is straightforward, although it may lead to some redundancy in the results.

Aside from constructing tables that summarize the relative performance of PV systems in different locations of Oman, this work also provides the year-round optimum orientation angles (azimuth and slope) for fixed PV panels in each analyzed location and the optimum slope angle for single-vertical-axis PV panels with solar tracking. The multiplication factors in the tables are all relative to the PV system electricity generation (alternating-current electricity) in Muscat, which is taken as a reference location and thus assigned the factors 1.00 by definition. This study also provides two normalized performance metrics for PV systems in each analyzed location: the first is the expected yearly PV electricity generated per kWp of installed PV capacity (expressed in kWh/kWp/year), and the second is the yearly global solar irradiation received per m^2 of tilted PV panel surface (expressed in kWh/m^2/year). The first normalized performance metric helps in evaluating the feasibility of investing in solar PV in Oman through estimating the size of the installed capacity needed. The second metric serves a similar role as the first one, but it is independent of the exact performance characteristics of the envisioned PV system. The work also provides some details about the solar PV electricity generation at the best-performing and the lowest-performing locations identified among the analyzed ones, as well as the reference location of Muscat, representing the capital of the country.

2. Methods

The study is primarily quantitative. It is primarily an analysis of secondary data and third-party validated solar PV web-based tools that are freely accessible to the public and are well documented. The tools are collectively referred to as the PVGIS (Photovoltaic Geographical Information System) web application, which includes a calculator for solar irradiation and PV electricity generation. It is hosted at the European Commission's science and knowledge service (EU Science Hub) and developed by the Joint Research Centre (JRC) of the European Commission [18]. The version used here is the 5th version, which is the latest version at the time of preparing this article. PVGIS has 5 solar radiation databases. However, the one that covers Oman and was used in the current study is PVGIS-SARAH. It is a data set calculated from satellite images, and covers the whole Sultanate of Oman, along with Africa, most of Europe, more than half of Asia, approximately the eastern half of South America, and a small zone in the west of Australia. The data in PVGIS-SARAH database cover the years from 2005 to 2016. Their spatial resolution is 0.05° (latitude) and 0.05° (longitude) [19], which, near the equator, translates to a grid of squares with a side length of about 5.5 km. The PVGIS-SARAH database was prepared through cooperation between members at PVGIS and at CM SAF (Satellite Application Facility on Climate Monitoring), which provides satellite-based data suitable to monitor the climate [20]. PVGIS was subject to data validation for the used satellite solar radiation data. Comparison with ground station measurements shows generally good agreement between the ground station data and the PVGIS-SARAH database [21–25], which is the one used in the current study for Oman. Most of the benchmarking data for validation came from the Baseline Surface Radiation Network (BSRN), which is a project for detecting change in solar and atmospheric radiation at the surface of the Earth, and that project belongs to the World Climate Research Programme (WCRP). The project's instruments are of the highest possible accuracy and provide high time resolution (from 1 to 3 min) [26].

The PVGIS application was utilized in calculating PV performance data for 40 locations in the Sultanate of Oman. When selecting these locations, attention was paid to cover a large geographic area of the country, while limiting the locations to the official second-level administrative divisions (wilayats). This is an advantage for the locations' selection, helping in aligning them with places where social communities typically exist.

Table 1 lists the wilayats selected for solar PV analysis. They are grouped by governorate, starting with the most-populous governorate (Muscat) and ending with the least-populous governorate (Musandam), based on recent governmental records dated August 2021 [27]. The number of wilayats in each governorate [28] and the number of selected wilayats for analysis in each governorate are also listed in the table. In total, 40 wilayats were selected out of the country-level total of 61 [29]. At minimum, either three or half of the number wilayats in each governorate is selected, whichever is larger. The governorates of Adh Dhahirah and Al Buraimi have 3 wilayats each, and thus all of them were included in the analysis. Using an official map of Oman [30] to inspect visually the separation distance among wilayats, some wilayats were excluded when found lying close to an already-selected wilayat. The first wilayat listed for each governorate is the administrative capital of that governorate. Then, when adding more wilayats, generally a more distant wilayat has priority of inclusion in the analysis, which helps in improving the geographic coverage of the study. The governorate of Dhofar is large in terms of the land area and is also relatively populous; it has the largest number of selected wilayats (seven) for solar PV analysis compared to other governorates. The governorate of Al Wusta is also large in terms of the land area, and all its 4 wilayats were selected in the analysis despite being the second least populated governate in Oman.

Table 1. List of the selected wilayats in the Sultanate of Oman for solar energy analysis.

Index of Selected Wilayats	Governorates (in Descending Order by Population)	Population of the Governorate (as of August 2021)	Number of Wilayats in the Governorate	Number of Selected Wilayats in the Governorate	Selected Wilayats for Analysis (First One is the Administrative Capital of the Governorate)
1 2 3	Muscat	1,270,308	6	3	Muscat Qurayyat A'Seeb
4 5 6	Al Batinah North	780,899	6	3	Sohar Al Suwaiq Shinas
7 8 9 10	Ad Dakhliya	478,229	8	4	Nizwa Adam Samail Al Hamra
11 12 13	Al Batinah South	465,210	6	3	Al Rustaq Barka Wadi Al Ma'awil
14 15 16 17 18 19 20	Dhofar	409,019	10	7	Salalah Muqshin Shalim and the Hallaniyat Islands Dhalkut Al Mazyona Thumrait Sadah
21 22 23	Ash Sharqiyah South	314,204	5	3	Sur Masirah Jaalan Bani Bu Ali
24 25 26 27	Ash Sharqiya North	270,520	6	4	Ibra Al Mudhaibi Wadi Bani Khalid Bidiya
28 29 30	Adh Dhahirah	212,954	3	3	Ibri Yankul Dhank
31 32 33	Al Buraimi	120,295	3	3	Al Buraimi Al Sunaynah Mahdah

Table 1. *Cont.*

Index of Selected Wilayats	Governorates (in Descending Order by Population)	Population of the Governorate (as of August 2021)	Number of Wilayats in the Governorate	Number of Selected Wilayats in the Governorate	Selected Wilayats for Analysis (First One is the Administrative Capital of the Governorate)
34 35 36 37	Al Wusta	51,191	4	4	Haima Duqm Al Jazer Mahout
38 39 40	Musandam	48,834	4	3	Khasab Dibba Madha
Total	-	4,421,663	61	40	-

More details about the selected locations are provided in Appendix A. These are the latitude and longitude coordinates, as well as the elevations.

3. Results

The performed PV solar simulations using the PVGIS tools assume crystalline silicon (c-Si) panels, which is the dominant technological type of PV panels [31–33], and a system loss of 14% (the default value in PVGIS). Such a system loss accounts for losses in cables and power inverters, for example, as well as any dirt on the panels [34]. The simulations performed for the design type of a fixed PV system with optimum orientation of the PV panels are for free-standing mounting, as compared to building integrated, so the virtual PV panels are supported by a frame allowing air to flow over the panels' back side.

Although the lookup tables are essential outcomes of this study, it may be better to move them to Appendix B, separated from the main text, while describing them in the present section. There are three lookup tables, each of them corresponding to a design type of the PV system.

3.1. Validation against Third-Party Calculated Results

This subsection presents a validation test for one of the solar calculator tools of PVGIS. This is different from the independent validation studies mentioned in the previous section. The validation is performed in terms of predicting the yearly electricity generation for 1 kWp of installed PV capacity, as well as the received solar radiation per 1 m^2 of the tilted PV panel surface in the location representing the capital Muscat, with a fixed optimized orientation of PV panels. These four simulation quantities are compared when obtained from PVGIS and from another online solar calculator, which is the Global Solar Atlas (GSA), a free tool provided by the World Bank Group [35].

The validation comparison is given in Table 2. The optimum slope (the tilt angle from the horizontal) suggested by PVGIS is only 1 degree smaller than that suggested by GSA. It should be mentioned that both tools report the slope angle in whole degrees. The angles are reasonable, as they should be close to (but not necessarily equal to) the latitude of the location [36] (which is 24° in the test case, after rounding to the nearest integer). Both tools agree on the suggested optimum azimuth angle of the PV panels in Muscat, showing that the PV panels should be facing the south exactly. This is a reasonable azimuth given that Muscat's selected analysis location is in the northern hemisphere [37], and it is not surrounded by excessive mountains that may cause some bias toward the west or the east. As a remark, the azimuth angle in PVGIS for the south is expressed as 0°, whereas it is expressed as 180° for GSA, given the different reference of measurement in both tools (the south for PVGIS, but the north for GSA). The yearly PV electricity generated for 1 kWp of PV capacity is very close for PVGIS and GSA, differing by only 10.2 kWh/kWp/year, which is 0.6% of either value. The estimated solar radiation differs by 123.3 kWh/m^2/year for the two tools, which is about 5% of either value.

Table 2. Comparison of PV system characteristics with optimized fixed orientation in the location representing Muscat (latitude: north 23.583889°, longitude: east 58.407778°), as predicted by PVGIS (Photovoltaic Geographical Information System) and by GSA (Global Solar Atlas) online solar calculation tools.

Characteristic	PVGIS Value	GSA Value	Absolute Difference
Optimum slope (°)	25	26	1
Optimum azimuth, from south (°)	0	0	0
Yearly PV electricity production per kWp installed (kWh/kWp/year)	1809.4	1819.6	10.2
Yearly global solar radiation received by m^2 of tilted PV panel surface (kWh/m^2/year)	2500.8	2377.5	123.3

3.2. PV Design Type 1: Fixed Optimized Orientation

Table A3 is the lookup table for the PV design type with fixed orientation where the photovoltaic panels are installed with the optimized azimuth angle (the direction in which the panels face in the horizontal plane) and the optimized slope angle (the tilt above the horizontal plane) for maximum cumulative yearly radiation received by a unit area of the photovoltaic panels, considering the whole year rather than a specific season. These optimized angels are reported as integer values by PVGIS.

For the analyzed locations, the optimized slope ranges from 20° (in multiple locations) to 26° (in Al Rustaq and in Dibba), with a range of 6°. In relation to this, the latitude angle for the analyzed locations varies from 16.70° (in Dhalkut) to 26.18° (in Khasab), with a range of 9.48°. Figure 2 shows how the slope changes with the latitude for all the analyzed locations. A linear regression model is moderately fitting the data.

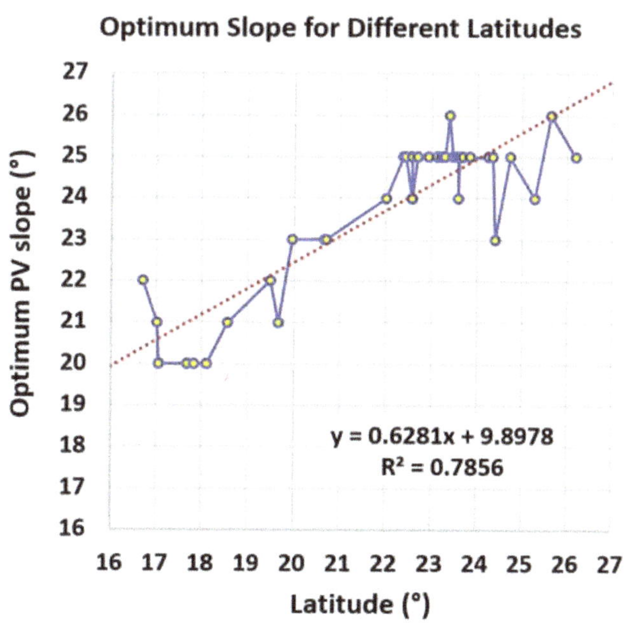

Figure 2. Variation of the optimized slope of PV panels (as predicted by PVGIS) with the latitude for the analyzed locations in the Sultanate Oman.

For the optimized PV azimuth, it is at or close to due south, with a small east–west deviation range of 16°. The most east-biased azimuth corresponds to the location of Madha (in Musandam governorate), being 11° toward the east. The most west-biased azimuth corresponds to the location of Salalah (in Dhofar governorate), being 5° toward the west. The profiles of the horizon height (expressed as an angular distance, in degrees) for both locations are compared in Figure 3. The deviation from due south in the case of Salalah is small and can be attributed to small differences in the solar radiation between the morning and the afternoon intervals.

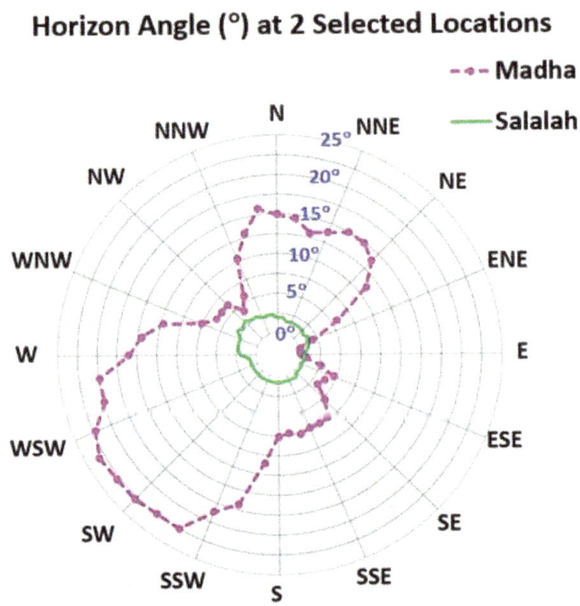

Figure 3. Illustration of the horizon height (in angular degrees) at two special locations in the Sultanate of Oman among Table 1. Madha (showing the most eastern-biased optimized azimuth angle for fixed PV panels, 11° east of due south); (2) Salalah (showing the most western-biased optimized azimuth angle for fixed PV panels, 5° west of due south). Data source: PVGIS ©European Union, 2001–2021. The innermost circle corresponds to a horizon height of 0°, and the outermost circle corresponds to a horizon height of 25°.

Figure 4 shows a calculated typical hourly profile of irradiance on the fixed optimally oriented PV panels in Salalah, calculated by one of the PVGIS online tools (AVERAGE DAILY IRRADIANCE DATA). The profile is not perfectly symmetric around the point of maximum irradiance, supporting the presence of a small difference in the cumulative received radiation before and after the solar noon. The deviation from due south in the case of Madha can be partly attributed to a relatively high horizon in the west part compared to the east part, leading to an early virtual sunset, with the sun no longer sending direct normal irradiance (DNI) to the PV panels from the west side, which makes the east side more favorable. In the case of Salalah, the yearly PV electricity production per kWp installed decreased from 1814.37 kWh/kWp/year to 1813.56 kWh/kWp/year (keeping the optimum slope as 21°), which is a trivial drop of less than 1 kWh. In the case of Madha, the yearly PV electricity production per kWp installed decreased from 1753.99 kWh/kWp/year to 1750.46 kWh/kWp/year (keeping the optimum slope as 24°). This is a small drop of only 0.2%.

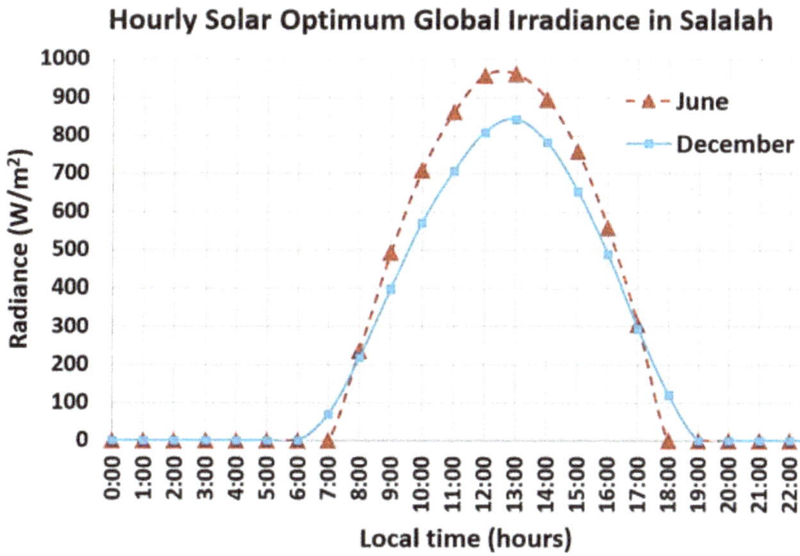

Figure 4. Temporal distribution of calculated hourly solar irradiance for the month of June and the month of December in the location of Salalah, latitude: north 17.015° and longitude: east 54.092222°. The values are obtained using PVGIS Table 2005. Years of data in the database: PVGIS-SARAH.

Concerning the normalized electricity generation (per kWp and year) for the analyzed locations, it ranges from 1753.99 kWh/kWp/year in Madha (in the northern exclave governorate of Musandam) to 1959.43 kWh/kWp/year in Thumrait (in the southern governorate of Dhofar), with a gap of 205.44 kWh/kWp/year between them. The average value of the normalized electricity generation over all analyzed locations is 1845.75 kWh/kWp/year.

3.3. PV Design Type 2: Single-Vertical-Axis Solar Tracking

The lookup Table A4 summarizes results for the PV design type with single-axis solar tracking, where the PV panels (which are tilted at an optimized slope) are able to rotate around a vertical axis. This means that the azimuth can change during the day, but the slope has to be fixed during that rotation.

The optimized slope in this type was found to vary over a narrow range from 46° to 49°. The slope is higher than it was with a fixed optimized azimuth (design type 1).

The normalized electricity generation for the analyzed locations ranges from 2181.92 kWh/kWp/year in Madha to 2559.34 kWh/kWp/year in Thumrait, with a gap of 377.42 kWh/kWp/year between them. The average value of the normalized electricity generation over all analyzed locations is 2373.68 kWh/kWp/year, which is 1.286 times its corresponding value for the fixed optimized azimuth (design type 1).

3.4. PV Design Type 3: Two-Axis Solar Tracking

The lookup Table A5 summarizes results for the PV design type with two-axis solar tracking, where the PV panels are able to rotate around both the vertical axis and the horizontal axis, thus receiving the solar rays perpendicularly all the time. This means that losses due to the imperfect orientation of PV panels are eliminated, since the PV orientation is continuously optimized [38].

The normalized electricity generation for the analyzed locations ranges from 2254.12 kWh/kWp/year in Madha to 2674.22 kWh/kWp/year in Thumrait, with a gap of 420.1 kWh/kWp/year between them. The average value of the normalized electricity generation over all analyzed locations is 2471.57 kWh/kWp/year, which is 1.339 times its corresponding value for the fixed optimized azimuth and 1.041 times its corresponding value for single-vertical-axis tracking.

3.5. Some Monthly Profiles of PV Electricity

The location (or wilayat) of Madha consistently gave the lowest normalized electricity generation among all the locations examined for the three design types of PV systems, whereas the location (or wilayat) of Thumrait consistently gave the highest normalized electricity generation. In addition, the improvement in the normalized electricity generation when upgrading from single-vertical-axis tracking to two-axis tracking was much smaller compared to upgrading from optimized fixed orientation to single-vertical-axis. Not only did the average normalized electricity increase when upgrading the PV design type, but the gap in the normalized electricity (difference between its maximum in Thumrait and its minimum in Madha) also increased.

Figures 5–7 give some insight about these findings, through examining the monthly profile of the normalized electricity generation for the three PV design types, while considering the location of highest performance (Thumrait) and the location of lowest performance (Madha). Muscat is also included in the figures, and it has an intermediate performance lying between the two extremum locations. The months of October–March contribute largely to the difference of yearly PV electricity generation among the three locations, whereas the monthly electricity generation during April–September is not significantly different among the three locations. In the months of October–March, the sun has relatively low altitudes, and a place like Madha with high horizon heights is adversely affected by less direct solar irradiation to PV panels, regardless of the presence of tracking. With no exception to the months of the year or locations of the three discussed here, the monthly gain due to upgrading from fixed optimized orientation to single-vertical-axis tracking is noticeably larger than the monthly gain when upgrading from single-vertical-axis tracking to two-axis tracking, with the ratios of the two monthly gains for the three locations and the 12 months (total 36 ratios) ranging between 3.45 and 11.39. Additionally, for each month in October–March, the gap in the monthly PV electricity generation between Madha and Thumrait is increased in either level of upgrade.

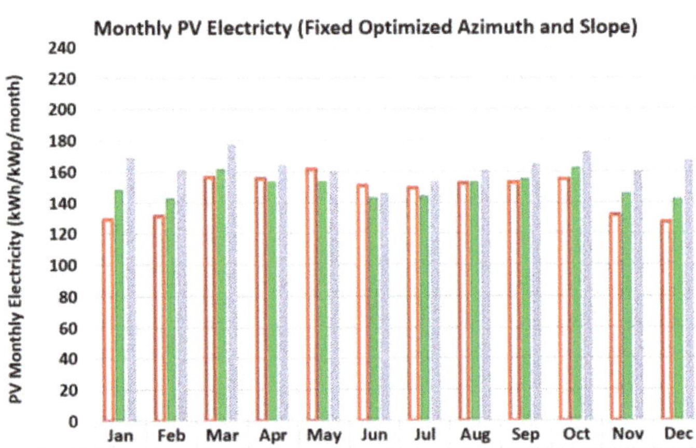

	Jan	Feb	Mar	Apr	May	Jun	Jul	Aug	Sep	Oct	Nov	Dec
Madha	129.3	131.3	156.6	155.6	161.5	150.9	149.3	152.2	153.1	155	131.7	127.3
Muscat	148.8	143.6	162.1	154	153.8	143.5	144.3	153.3	155.5	162.4	146.1	142.1
Thumrait	169.3	161.6	177.6	164.4	160.4	146.6	153.7	161.2	165.1	172.7	160	166.9

Figure 5. Monthly expected electricity generation from solar PV systems with fixed year-round optimized orientation of Table 1. (1) Madha (showing the lowest expected electricity generation over the whole year); (2) Muscat (the capital, and the reference location in the current study); (3) Thumrait (showing the highest expected electricity generation over the whole year). Data source: PVGIS ©European Union, 2001–2021.

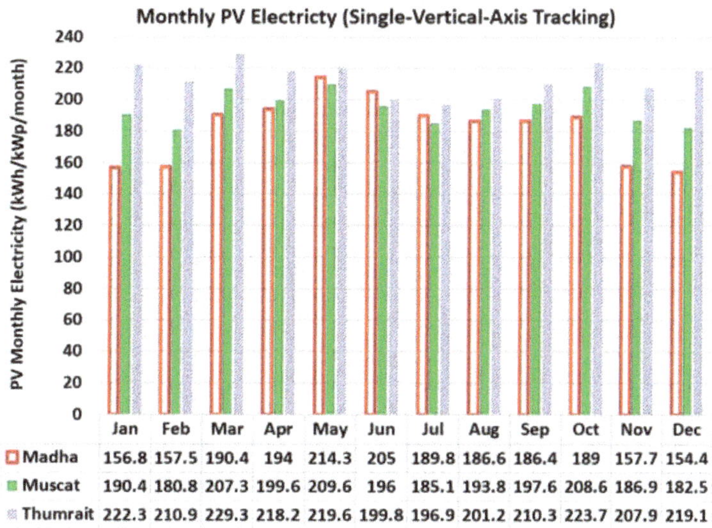

Figure 6. Monthly expected electricity generation from solar PV systems with single-vertical-axis solar tracking in three special locations in the Sultanate of Oman among those analyzed in the current study: (1) Madha (showing the lowest expected electricity generation over the whole year); (2) Muscat (the capital, and the reference location in the current study); (3) Thumrait (showing the highest expected electricity generation over the whole year). Data source: PVGIS ©European Union, 2001–2021.

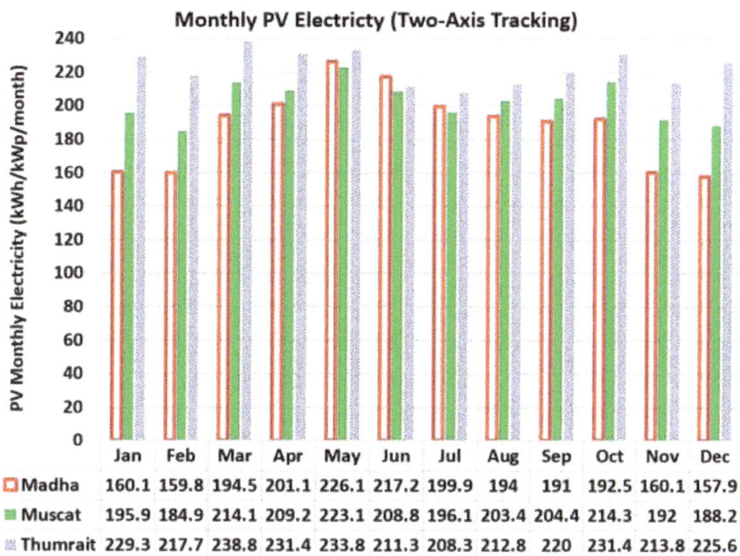

Figure 7. Monthly expected electricity generation from solar PV systems with two-axis solar tracking in three special locations in the Sultanate of Oman among those analyzed in the current study: (1) Madha (showing the lowest expected electricity generation over the whole year); (2) Muscat (the capital, and the reference location in the current study); (3) Thumrait (showing the highest expected electricity generation over the whole year). Data source: PVGIS ©European Union, 2001–2021.

3.6. General Perspectives and Remarks about Contributions

This work has a focus on the Sultane of Oman, establishing a numerical multi-aspect discrete-point map of its photovoltaic-based electricity generation potential, which may be considered as a local contribution at the country level. This is different from continuous two-dimensional maps [39] that are excellent in quickly showing spatial variations and zones of peak performance, but not as accurate and useful as lookup tables when a designer is in need of precise numbers at a precise point. While the raw data itself used in the work are not novel, their analytics (collection, organization, processing, checking, visualization) results in a novel output for the local reader.

Considering global readers, they may still benefit from the earlier part through exposure to web-based modeling tools and their capabilities (in case they are not already familiar with them) and through the suggested relation between the optimum slope for fixed PV panels and the latitude. Additionally, the asymmetry in solar irradiance during the day and the impact of horizon discussed here bring specific examples of features influencing solar-power performance that are applicable globally.

In the current subsection, the non-local contribution of this work is strengthened in two ways: first, through presenting a conducted validation of PVGIS calculations against monthly ground-measured global horizontal irradiation (GHI), which also tests the ability of the PVGIS tool in capturing weather disturbances.

As with the PVGIS data, the measurements used here are not proprietary. They were published openly by the Oman Power and Water Procurement Company (OPWP) [40], which is the sole entity involved in purchasing electricity and water for all independent power projects (IPPs) or independent water and power projects (IWPPs) in the Sultanate of Oman [41]. OPWP is in charge of ensuring that the needs of electricity and water in the Sultanate of Oman are met, with adequate supply of electricity and water, at competitive prices. OPWP is part of the Nama Holding Company (previously called Electricity Holding Company), which is a joint stock governmental company, having a share in multiple companies operating collectively in the fields of the procurement, generation, transmission, and distribution of electricity and water within the Sultanate of Oman [42,43].

The experimental study reports the measured average daily global horizontal irradiation (GHI), expressed in kWh/m^2/day for 11 months in 2013 (all months in that year except the month of January). Here, the months of February, May, and June were excluded despite having reported data. The reason is that the data sets for these months are not complete (some days of the month do not have collected data). The remaining 8 months in 2013 with complete GHI solar irradiation measurements are March, April, July, August, September, October, November, and December. The data used here come from the Adam meteorological station. Adam is one of the 40 wilayats analyzed here in terms of PV performance (in the governorate of Ad Dakhliya). The experimental daily GHI measurement is converted into a monthly value by multiplying it by the number of days in the month (either 30 or 31). This makes the unit of the resulting number kWh/m^2/month, which is consistent with a tool in PVGIS used to calculate the GHI at the location representing Adam here (latitude: north 22.379167° N, longitude: east 57.526944° E). The PVGIS tool used here is called (MONTHLY IRRADIATION DATA), and it reports monthly GHI in kWh/m^2/month. The PVGIS data correspond to the same year of the measurements, which is 2013.

The experimental data are based on true measurements, which reflect weather irregularities (overcast and rainfall). Corrections were also reported in the third-party experimental study of OPWP to enable the construction of fictitious GHI values assuming nearly clear days. The correction is performed here, and the adjusted experimental monthly GHI values were obtained. Such corrections are not applicable for the months of September and October, because no important weather irregularities were observed and the true measured GHI and the adjusted measured GHI are considered the same.

The percentage difference between the PVGIS-based GHI and the experimental GHI is calculated as in Equation (1):

$$\text{Percentage Difference} = \frac{\text{PVGIS Value} - \text{Meaured Value}}{\text{Measured Value}} \times 100\%. \quad (1)$$

With respect to the true measurements, all the individual monthly percentage differences are positive, which means that the calculated values of GHI exceed the experimental GHI values. Nevertheless, considering all 8 months, the average of the percentage differences is 3.32%, which is relatively small. With respect to the adjusted measurements, half of the individual monthly percentage differences are positive, and the other half is negative. The average of the absolute values of the percentage difference for the 8 months is 2.82%. The individual monthly percentage differences are shown in Figure 8.

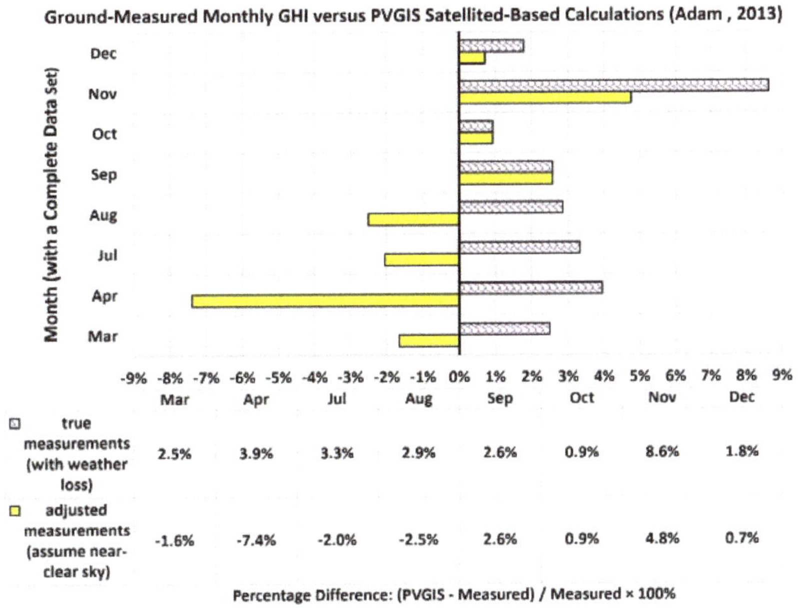

Figure 8. Comparison of monthly global horizontal irradiation (GHI) for the location of Adam, located in the governorate of Ad Dakhliya, in the Sultanate of Oman, when measured at a meteorological station and when reported by PVGIS web calculator (PVGIS ©European Union, 2001–2021), for 8 months in 2013.

As another component in the present work that goes beyond the local scope of Oman and can be beneficial for global readers, the current part examines the longitude-dependence of solar irradiation (while having the latitude nearly fixed, thus having similar distances from the equator) on a large geographic scale. This component is conducted by selecting additional locations outside Oman with a latitude lying between 23° and 25° (regardless of being in the northern hemisphere or the southern hemisphere), similar to Muscat in Oman, and comparing the annual global horizontal irradiation (GHI) in $MWh/m^2/year$ and the annual direct normal irradiation (DNI) in $MWh/m^2/year$ among them, as predicted by PVGIS (obtained by summing the 12 monthly values in the same year, from January to December). A large variability calls for attention among solar energy modelers who may rely on simple equations for incoming annual solar radiation that heavily depends on the latitude (while not taking into account other secondary factors, such as the terrain, cloud cover, and horizon heights).

In addition to Muscat (the Capital of Oman and one of the 40 local locations analyzed here), the additional foreign locations are distributed in different countries (one location per country) in North America, South America, Africa, Asia, and Australia. Out of the total of 12 international locations, three are in southern hemisphere (such as Gaborone, the capital of Botswana in Africa) and the remaining nine locations are in the northern hemisphere (such as Culiacán, a city in northwestern Mexico, and Dhaka, the capital of Bangladesh). More details about these 12 locations (the latitude, the longitude, and the elevations) are provided in Table A2 in Appendix A.

Nearly all the additional foreign locations were analyzed using the PVGIS-SARAH data set (which is the data set used for the 40 wilayats in Oman in the earlier locally focused part of this work) except Culiacán (Mexico) and Havana (the capital of Cuba), where PVGIS-SARAH is not applicable. Instead, the data set of PVGIS-NSRDB was used. PVGIS-NSRDB has a finer spatial resolution than PVGIS-SARAH (latitude × longitude: 0.038° × 0.0.38°, compared to 0.05° × 0.05° for PVGIS-SARAH), but its end year is 2015 (as compared to 2016 for PVGIS-SARAH). The calculated annual GHI and DNI here are for the last year available in the data set, which is either 2016 for PVGIS-SARAH or 2015 for PVGIS-NSRDB.

Figure 9 presents the annual GHI for the 12 locations in the selected 12 countries, ordered from smallest (top) to largest (bottom). The largest-to-smallest GHI ratio is 1.64. Al Jawf in Libya (North Africa) has the largest GHI, which is slightly above the annual GHI of Muscat (with a ratio of 1.088). Muscat comes in the third position when ordering the GHI in a descending order.

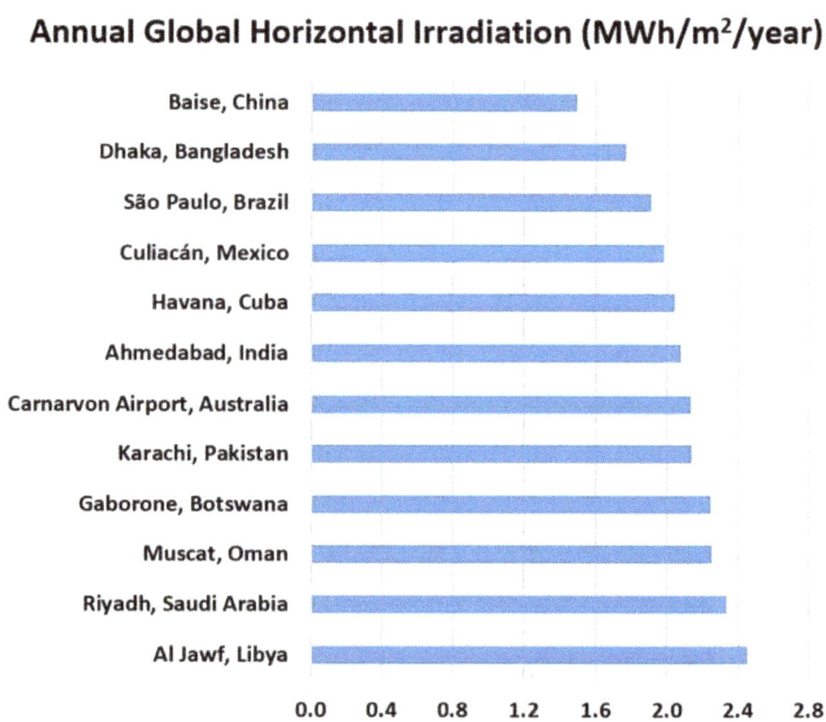

Figure 9. Annual global horizontal irradiation (GHI) for 12 locations in 12 countries having similar distances from the equator as Muscat. Data source: PVGIS ©European Union, 2001–2021.

Figure 10 presents the annual DNI for the same 12 locations selected in 12 countries, following the same order as in the previous figure (ascending order of GHI). The largest-to-smallest DNI ratio is 2.77, which is significant and reflects a more intensive variability in

DNI than in GHI. Al Jawf in Libya has the largest DNI (as was the case for GHI), which is 1.157 times the DNI of Muscat. Muscat comes in the fifth position when ordering the DNI in a descending order.

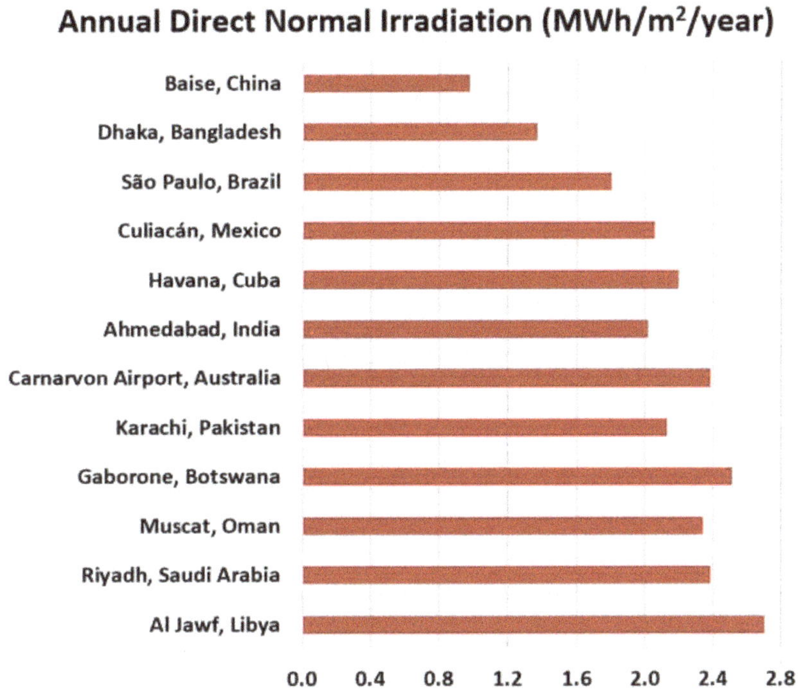

Figure 10. Annual direct normal irradiation (DNI) for 12 locations in 12 countries having similar distances from the equator as Muscat. Data source: PVGIS ©European Union, 2001–2021.

4. Discussion

This section gives some remarks about the results of the study. The first remark concerns the results' generality. For the presented suggested optimized slope, optimized azimuth, and normalized solar irradiation, they should be stable in terms of not being influenced by the selected simulation parameter. On the other hand, the normalized PV electricity generation corresponds to a given mounting type and a specified system loss. In the case of the Muscat location, the building-integrated mounting was predicted to result in a lower normalized PV electricity generation of 1725.1 kWh/kWp/year as compared to 1809.44 kWh/kWp/year under free-standing mounting with fixed optimized PV orientation, which means a drop of 84.34 kWh/kWp/year (thus 4.66% relative to the free-standing case). If the mounting type changes from free standing to building integrated, the normalized PV electricity generation drops from 1753.99 kWh/kWp/year to 1675.79 kWh/kWp/year for Madha (a drop of 78.20 kWh/kWp/year or 4.46% relative to the free-standing case) and from 1959.43 kWh/kWp/year to 1874.35 kWh/kWp/year for Thumrait (a drop of 85.08 kWh/kWp/year or 4.34% relative to the free-standing case). Based on these cases, deducting 80 kWh/kWp/year from the free-standing normalized electricity generation appears to be a reasonable way to estimate the building-integrated normalized electricity generation. This drop in normalized PV electricity generation can be justified by the lower cooling of the PV panels in the case of building-integrated mounting, thereby increasing the panels' temperature and decreasing their ability to generate electricity [44]. The influence of mounting and preset system loss is relatively mitigated by the relative performance multiplication factors provided in the lookup tables, where the PV electricity generation in a location is expressed in relation to its (hypothetical) counterpart in Muscat. This is because the PV electricity generation in both conditions was predicted under the same simulation settings.

The second remark is about the use of the relative-performance multiplication factors, relating the solar PV electricity generation in a general analyzed location to the reference one of Muscat. While any location among those analyzed can serve as a reference for others, Muscat is the capital of the Sultanate of Oman, and it is a good candidate for setting a common base value. One multiplication factor is provided in the lookup tables with the name Translation Factor (X to Muscat), which manifests its role in translating or projecting the solar PV electricity generation from one location to another. The letter "X" refers to any location among the 40 analyzed ones in the Sultanate of Oman. The factor (X to Muscat) is a scaling value assigned to one location, such that when it is multiplied by the PV solar electricity in that location, one obtains an expected PV solar electricity in Muscat for the same PV system. Dividing by that factor serves the opposite function: scaling the yearly electricity generation in Muscat to its expected value in case the same PV system is translated from Muscat to another location "X".

For example, if a solar PV system with a fixed optimized PV orientation is located in Madha and was found to generate 800 kWh of electricity yearly, and then it is translated to Muscat, it is expected to produce 800 × 1.0316 = 825.28 kWh yearly. The number (1.0316) is the translation factor (X to Muscat) for Madha in the lookup table under the fixed optimized-orientation PV design type. As another example, if a solar PV system with a fixed optimized orientation is located in Muscat and was found to generate 800 kWh of electricity yearly, and then it is translated to Thumrait, it is expected to produce 800 ÷ 0.92345 = 866.32 kWh yearly. The number (0.92345) is the translation factor (X to Muscat) for Thumrait under the fixed optimized-orientation PV design type. Translating the yearly electricity generation from one location to another when both locations are not Muscat involves a multiplication with the (X to Muscat) translation factor for the source location combined with a division by the (X to Muscat) translation factor for the destination location. For example, if a solar PV system with a fixed optimized orientation is located in Madha and was found to generate 800 kWh of electricity yearly, and then it is translated to Thumrait, it is expected to produce 800 × 1.0316 ÷ 0.92345 = 893.7 kWh yearly. The numbers (1.0316) and (0.92345) are interpreted here exactly as they were interpreted earlier. The translation (or projection) process is not limited to a system size (no need to restrain them to 1 kWp of installed PV capacity) but should be applied to a one-year interval (to adjust yearly electricity generation).

The largest relative deviation in the predicted yearly PV electricity generation due to translating a PV system from one place to another among those analyzed in the present study is simply obtained through dividing the largest (X to Muscat) translation factor by the smallest (X to Muscat) translation factor in the same lookup table. As a special case in the present study, the largest (X to Muscat) translation factor is always the one corresponding to the wilayat of Madha, while the smallest (X to Muscat) translation factor is always the one corresponding to the wilayat of Thumrait. The largest relative deviations are as follows:

- For fixed optimized orientation: 1.0316 ÷ 0.92345 = 1.117 (or 11.17% increase);
- For single-vertical-axis tracking: 1.0716 ÷ 0.91358 = 1.173 (or 11.73% increase);
- For two-axis tracking: 1.0800 ÷ 0.91032 = 1.186 (or 11.86% increase).

These extreme-case values are not very large, suggesting that the location alone is not a key element in improving solar PV generation in the Sultanate of Oman. Considering all 3 PV design types and all 40 analyzed locations (thus, a total of 120 cases), the highest predicted normalized PV electricity generation is that for Thumrait with two-axis solar tracking, which is 2674.22 kWh/kWp/year, while the lowest predicted normalized PV electricity generation is that for Madha with fixed PV optimized orientation, which is 1753.99 kWh/kWp/year. The ratio of the two overall extreme values is 1.5246.

As a third remark, while the current study pays special attention to the Sultanate of Oman in terms of lookup tables and graphical results, the procedure of establishing these results is not restrained to a specific country or region. In addition, the tools needed to build the lookup tables for certain locations of interest are made available to the public and involve a validated solar simulation application. Thus, the present study can be useful to both a local audience in the Sultanate of Oman and an overseas audience outside.

5. Conclusions

This study may be viewed as a collection of data related to the expected performance of solar photovoltaic (PV) systems in the Sultanate of Oman, providing quick estimates of normalized input irradiation (in kWh of radiative energy for 1 m^2 of tilted photovoltaic panels area and over one year) and normalized output electricity generation (in kWh of electric energy for 1 kWp of installed photovoltaic power and over one year) in more than half of the second-level administrative divisions (wilayats) of the country. The reader does not need to have technical background in solar simulation and does not need to perform complicated data analysis. The results are normalized, making them easy to scale to applications of specific size through one simple arithmetic operation. Additionally, the PV electricity generation in one location can be projected to another. In addition, the optimum slope and azimuth for fixed PV panels are given for each location.

Three lookup tables were established, corresponding to three design types of PV panels, being (1) fixed optimized slope and azimuth, (2) single-vertical-axis solar tracking, and (3) two-axis solar tracking. The upgrade from the fixed optimum slope and azimuth to the single-vertical-axis tracking has much more improvement than the upgrade from the single-vertical-axis tracking to the two-axis tracking.

The location of Madha (an exclave zone in the north of Sultanate of Oman) was found to have the least expected electricity generation, whereas the location of Thumrait (in the south of Sultanate of Oman) was found to have the best expected electricity generation. These rankings for both locations are irrespective of the PV design type. The monthly electricity generation was examined for both locations.

The study focused on free-standing PV mounting. However, the study indicates that it can be enough to subtract 80 kWh/kWp/year from the normalized PV electricity generation for the PV design type of fixed optimized orientation to reach an estimate for building-integrated PV mounting.

A brief benchmarking test was conducted, comparing PVGIS to GSA (Global Solar Atlas) in the case of the fixed optimized slope and azimuth for Muscat.

When comparing PVGIS calculations of monthly global horizontal irradiation (GHI) for 8 months in 2013 with ground-measured values in a meteorological station in Oman, reasonable performance was observed overall. The percentage differences (with the measurements taken as a reference) ranged from 0.9% (October) to 8.6% (November).

The GHI and the direct normal irradiation (DNI) calculation capability in PVGIS were used to estimate the variability in annual solar irradiation with the longitude (while keeping the latitude within a narrow zone of only 2 degrees). Moderate variation was found in the GHI, but remarkable deviations were found for the DNI among 12 locations (including Muscat) in 12 countries that were analyzed. This step shows the importance of using detailed solar energy modeling tools and not relying on simplified models when predicting the performance of a solar system at a large commercial scale.

Funding: This research received no external funding.

Institutional Review Board Statement: Not applicable.

Informed Consent Statement: Not applicable.

Data Availability Statement: Publicly available datasets were analyzed in this study. These data may be found in links provided in relevant references in this article, such as [35,40,45,46].

Acknowledgments: The National Centre for Statistics and Information of the Sultanate of Oman is highly thanked for quickly (after only one hour) clarifying that no permission is needed from them to use their maps, which were used to construct the map included in the study for the Sultanate of Oman, forming an important component of this study, especially for those not very familiar with the country.

Conflicts of Interest: The author declare no conflict of interest.

Appendix A

The first part of this appendix is devoted to giving some geographic information about the 40 selected local locations (wilayats) in the Sultanate of Oman for the solar PV analysis in the current study. The second part of this appendix gives a similar set of information about the 12 selected international locations in 12 different countries in the world (including the Sultanate of Oman), to examine the variability in solar irradiation even with similar latitudes.

For each location, the latitudes and longitudes [45] are listed in the degrees, arcminutes, arcseconds format (DDMMSS), and the elevations are also given in meters [46]. Each latitude/longitude coordinate pair was verified to lie in the corresponding wilayat using Google Maps [47].

Table A1. Geographic information about the locations selected for solar PV analysis in the Sultanate of Oman.

Name of Selected Wilayat	Latitude (Degrees, Arcminutes, Arcseconds)	Longitude (Degrees, Arcminutes, Arcseconds)	Elevation (m)
Muscat	N 23°35′2″	E 58°24′28″	17
Qurayyat	N 23°15′46″	E 58°55′12″	7
A'Seeb	N 23°40′13″	E 58°11′20″	6
Sohar	N 24°20′50″	E 56°42′33″	11
Al Suwaiq	N 23°50′58″	E 57°26′19″	6
Shinas	N 24°44′33″	E 56°28′1″	2
Nizwa	N 22°56′0″	E 57°32′0″	507
Adam	N 22°22′45″	E 57°31′37″	284
Samail	N 23°17′54″	E 57°57′11″	402
Al Hamra	N 23°6′54″	E 57°17′29″	649
Al Rustaq	N 23°23′27″	E 57°25′28′	344
Barka	N 23°40′43″	E 57°53′9″	13
Wadi Al Ma'awil	N 23°27′51″	E 57°49′18″	212
Salalah	N 17°0′54″	E 54°5′32″	16
Muqshin	N 19°29′53″	E 54°50′36″	113
Shalim and the Hallaniyat Islands	N 18°6′9″	E 55°39′12″	269
Dhalkut	N 16°42′14″	E 53°14′31″	9
Al Mazyona	N 17°50′18″	E 52°39′37″	504
Thumrait	N 17°40′12″	E 54°2′0″	454
Sadah	N 17°3′0″	E 55°4′0″	44
Sur	N 22°34′0″	E 59°31′44″	7
Masirah	N 20°39′2″	E 58°52′22″	12
Jaalan Bani Bu Ali	N 22°0′56″	E 59°20′1″	92
Ibra	N 22°41′26″	E 58°32′0″	436
Al Mudhaibi	N 22°34′13″	E 58°7′16″	401
Wadi Bani Khalid	N 22°34′48″	E 59°5′35″	600
Bidiya	N 22°26′25″	E 58°47′57″	290
Ibri	N 23°13′32″	E 56°30′56″	361
Yankul	N 23°35′11″	E 56°32′22″	546
Dhank	N 23°33′1″	E 56°15′26″	342
Al Buraimi	N 24°15′3″	E 55°47′35″	298
Al Sunaynah	N 23°36′23″	E 55°57′16″	255
Mahdah	N 24°24′1″	E 55°58′1″	426
Haima	N 19°57′33″	E 56°16′32″	135
Duqm	N 19°39′43″	E 57°42′13″	4
Al Jazer	N 18°32′42″	E 56°21′36″	144
Mahout	N 20°42′0″	E 57°55′19″	40
Khasab	N 26°10′47″	E 56°14′51″	11
Dibba	N 25°38′11″	E 56°15′57″	1
Madha	N 25°16′53″	E 56°18′57″	97

Table A2. Summary of location from different countries with similar distances from the equator.

Location	Country	Latitude (Degrees, Arcminutes, Arcseconds)	Longitude (Degrees, Arcminutes, Arcseconds)	Elevation (m)
Al Jawf	Libya	N 24°11′56″	E 23°17′27″	391
Riyadh	Saudi Arabia	N 24°41′15″	E 46°43′18″	612
Muscat	Oman	N 23°35′2″	E 58°24′28″	17
Gaborone	Botswana	S 24°39′16″	E 25°54′30″	1013
Karachi	Pakistan	N 24°51′38″	E 67°0′37″	10
Carnarvon Airport	Australia	S 24°52′50″	E 113°40′19″	5
Ahmedabad	India	N 23°1′32″	E 72°35′14″	55
Havana	Cuba	N 23°7′58″	W 82°22′58″	37
Culiacán	Mexico	N 24°47′25″	W 107°23′16″	92
São Paulo	Brazil	S 23°32′51′	W 46°38′10″	784
Dhaka	Bangladesh	N 23°42′37″	E 90°24′26″	17
Baise	China	N 23°53′59″	E 106°36′48″	134

Appendix B

This appendix provides three tables (lookup tables), which are summaries of normalized photovoltaic electricity generation in the selected locations (wilayats) in the Sultanate of Oman, both as absolute measures (in kWh/kWp/year) and as relative measures, relative to the location in the capital Muscat (nondimensional). Each row in either table corresponds to a location, where some characteristics related to solar PV performance are given. There are three tables corresponding to three different design types of the PV panels in terms of their response to the sun path.

Table A3. Lookup table for solar PV systems with fixed optimized orientation of the photovoltaic panels. The reference (yearly PV electricity production per kWp installed) is 1809.44 kWh/kWp/year, corresponding to the wilayat of Muscat.

Name of Selected Wilayat	Optimized Slope (°)	Optimized Azimuth, from South (°)	Yearly Global Solar Irradiation Received by m^2 of Tilted PV Panels Surface (kWh/m^2/year)	Translation Factor (X to Muscat) for Yearly PV Electricity Production
Muscat	25	0	2500.82	1.0000
Qurayyat	25	0	2489.68	0.99495
A'Seeb	25	0	2502.94	0.99855
Sohar	25	0	2509.63	0.99392
Al Suwaiq	25	0	2515.81	0.98945
Shinas	25	1	2493.46	0.98423
Nizwa	25	−4	2516.22	0.99362
Adam	25	−3	2575.14	0.96714
Samail	25	−7	2492.28	1.0004
Al Hamra	25	−7	2507.2	0.99871
Al Rustaq	26	0	2469.7	1.0116
Barka	25	0	2506.1	0.99854
Wadi Al Ma'awil	25	−2	2496.78	1.0040
Salalah	21	5	2383.35	0.99728
Muqshin	22	−1	2583.42	0.95280
Shalim and the Hallaniyat Islands	20	0	2562.06	0.94005
Dhalkut	22	0	2328.4	1.0246
Al Mazyona	20	−4	2610.95	0.94111
Thumrait	20	0	2618.37	0.92345
Sadah	20	2	2440.66	0.96510
Sur	24	0	2523.19	0.98118
Masirah	23	4	2514.08	0.94424
Jaalan Bani Bu Ali	24	−1	2552.07	0.96129
Ibra	25	−4	2522.88	0.98923
Al Mudhaibi	25	−4	2535.23	0.98424
Wadi Bani Khalid	24	−5	2505.25	0.98250
Bidiya	25	−2	2564.14	0.96101
Ibri	25	−5	2547.62	0.98085
Yankul	24	−2	2487.99	1.0019
Dhank	25	−5	2539.44	0.98800
Al Buraimi	25	−4	2499.83	0.99104
Al Sunaynah	25	−4	2542.3	0.97983
Mahdah	23	−1	2582.44	0.94764
Haima	23	−1	2582.44	0.94764
Duqm	21	1	2458.12	0.96761
Al Jazer	21	1	2582.31	0.93129
Mahout	23	−1	2551.41	0.95471
Khasab	25	3	2410.21	1.0069
Dibba	26	0	2384.36	1.0294
Madha	24	−11	2355.45	1.0316

Table A4. Lookup table for solar PV systems with single-vertical-axis solar tracking of the photovoltaic panels. The reference (yearly PV electricity production per kWp installed) is 2338.16 kWh/kWp/year, corresponding to the wilayat of Muscat.

Name of Selected Wilayat	Optimized Slope (°)	Yearly Global Solar Radiation Received by m^2 of Tilted PV Panels Surface (kWh/m^2/year)	Translation Factor (X to Muscat) for Yearly PV Electricity Production
Muscat	49	3209.54	1.0000
Qurayyat	49	3192.19	0.99431
A'Seeb	49	3216.3	0.99672
Sohar	49	3220.74	0.99354
Al Suwaiq	49	3232.53	0.98801
Shinas	49	3188.7	0.98879
Nizwa	48	3203.7	1.0040
Adam	49	3304.69	0.96816
Samail	48	3171.57	1.0105
Al Hamra	48	3191.02	1.0095
Al Rustaq	48	3136.45	1.0247
Barka	49	3221.28	0.99659
Wadi Al Ma'awil	49	3201.32	1.0062
Salalah	47	3032.6	1.0071
Muqshin	48	3315.31	0.95239
Shalim and the Hallaniyat Islands	47	3307.1	0.93211
Dhalkut	47	2945.66	1.0409
Al Mazyona	47	3373.23	0.93417
Thumrait	47	3390.48	0.91358
Sadah	46	3111.59	0.97085
Sur	48	3231.73	0.98079
Masirah	47	3201.48	0.95063
Jaalan Bani Bu Ali	48	3268.93	0.96041
Ibra	49	3229.52	0.99186
Al Mudhaibi	48	3248.05	0.98682
Wadi Bani Khalid	47	3160.65	1.0034
Bidiya	48	3282.28	0.96358
Ibri	49	3263.67	0.98308
Yankul	47	3136.6	1.0236
Dhank	49	3252.63	0.99104
Al Buraimi	49	3207.5	0.99030
Al Sunaynah	49	3263.15	0.97923
Mahdah	48	3304.5	0.95005
Haima	48	3304.5	0.95005
Duqm	47	3115.44	0.97770
Al Jazer	47	3319.9	0.92742
Mahout	48	3261.77	0.95623
Khasab	47	3007.45	1.0421
Dibba	49	3044.7	1.0339
Madha	46	2923.53	1.0716

Table A5. Lookup table for solar PV systems with two-axis solar tracking of the photovoltaic panels. The reference (yearly PV electricity production per kWp installed) is 2434.4 kWh/kWp/year, corresponding to the wilayat of Muscat.

Name of Selected Wilayat	Yearly Global Solar Radiation Received by m^2 of Tilted PV Panels Surface ($kWh/m^2/year$)	Translation Factor (X to Muscat) for Yearly PV Electricity Production
Muscat	3351.52	1.0000
Qurayyat	3334.6	0.99371
A'Seeb	3359.41	0.99634
Sohar	3362.95	0.99357
Al Suwaiq	3376.56	0.98769
Shinas	3326.57	0.98946
Nizwa	3342.55	1.0053
Adam	3452.65	0.96774
Samail	3307.15	1.0123
Al Hamra	3332.27	1.0097
Al Rustaq	3268.32	1.0273
Barka	3364.62	0.99632
Wadi Al Ma'awil	3342.71	1.0067
Salalah	3173.52	1.0037
Muqshin	3464.9	0.95139
Shalim and the Hallaniyat Islands	3461.08	0.92908
Dhalkut	3080.14	1.0385
Al Mazyona	3532.27	0.93145
Thumrait	3550.34	0.91032
Sadah	3254.42	0.96772
Sur	3375.57	0.98024
Masirah	3344.8	0.94891
Jaalan Bani Bu Ali	3415.83	0.95906
Ibra	3374.9	0.99139
Al Mudhaibi	3394.07	0.98633
Wadi Bani Khalid	3288.11	1.0074
Bidiya	3428.77	0.96304
Ibri	3409.06	0.98285
Yankul	3262.4	1.0280
Dhank	3396.47	0.99133
Al Buraimi	3348.81	0.99039
Al Sunaynah	3407.74	0.97915
Mahdah	3451.71	0.94934
Haima	3451.71	0.94934
Duqm	3254.41	0.97606
Al Jazer	3471.93	0.92503
Mahout	3407.27	0.95505
Khasab	3111.62	1.0517
Dibba	3174.02	1.0352
Madha	3027.54	1.0800

References

1. Masson-Delmotte, V.; Zhai, P.; Pörtner, H.-O.; Roberts, D.; Skea, J.; Shukla, P.R.; Pirani, A.; Moufouma-Okia, W.; Péan, C.; Pidcock, R.; et al. (Eds.) IPCC, 2018: Summary for Policymakers. In *Global Warming of 1.5 °C. An IPCC Special Report on the Impacts of Global Warming of 1.5 °C Above Pre-Industrial Levels and Related Global Greenhouse Gas Emission Pathways, in the Context of Strengthening the Global Response to the Threat of Climate Change, Sustainable Development, and Efforts to Eradicate Poverty*; World Meteorological Organization: Geneva, Switzerland, 2018; p. 32. Available online: https://www.ipcc.ch/site/assets/uploads/sites/2/2019/05/SR15_SPM_version_stand_alone_HR.pdf (accessed on 18 September 2021).

2. IRENA (International Renewable Energy Agency). *World Energy Transitions Outlook: 1.5 °C Pathway*; IRENA: Abu Dhabi, United Arab Emirates, 2021; p. 321. Available online: https://www.irena.org/publications/2021/Jun/-/media/C3BB1085FEA5401D9CA9598A61C8B267.ashx (accessed on 29 July 2021).
3. Distribution of Renewable Energy Technologies by IRENA (International Renewable Energy Agency). Available online: https://www.irena.org/Statistics/View-Data-by-Topic/Capacity-and-Generation/Technologie (accessed on 25 September 2021).
4. Remap—Renewable Energy Roadmaps by IRENA (International Renewable Energy Agency). Available online: https://www.irena.org/remap (accessed on 25 September 2021).
5. Solar PV Power Generation in the Sustainable Development Scenario, 2000–2030. IEA (International Energy Agency), Paris. Available online: https://www.iea.org/data-and-statistics/charts/solar-pv-power-generation-in-the-sustainable-development-scenario-2000-2030 (accessed on 25 September 2021).
6. Alboteanu, I.L.; Bulucea, C.A.; Degeratu, S. Estimating Solar Irradiation Absorbed by Photovoltaic Panels with Low Concentration Located in Craiova, Romania. *Sustainability* **2015**, *7*, 2644–2661. [CrossRef]
7. Katiyar, A.K.; Pandey, C.K. A Review of Solar Radiation Models—Part I. *J. Renew. Energy* **2013**, *2013*, 11. [CrossRef]
8. Bakirci, K. Models for Determination of Solar Energy Potential. *Energy Explor. Exploit.* **2008**, *26*, 281–292. [CrossRef]
9. Xie, C.; Schimpf, C.; Chao, J.; Nourian, S.; Massicotte, J. Learning and teaching engineering design through modeling and simulation on a CAD platform. *Comput. Appl. Eng. Educ.* **2018**, *26*, 824–840. [CrossRef]
10. Sharma, S.; Kurian, C.P.; Paragond, L.S. Solar PV System Design Using PVsyst: A Case Study of an Academic Institute. In Proceedings of the International Conference on Control, Power, Communication and Computing Technologies (ICCPCCT), Kannur, India, 23–24 March 2018; IEEE (Institute of Electrical and Electronics Engineers): New York City, NY, USA; pp. 123–128. [CrossRef]
11. Sources of Meteo Data in Monthly Values—PVsyst. Available online: https://www.pvsyst.com/meteo-data-source (accessed on 25 September 2021).
12. Honnurvali, M.S.; Gupta, N.; Goh, K.; Umar, T.; Kabbani, A.; Nazeema, N. Case study of PV output power degradation rates in Oman. *IET Renew. Power Gener.* **2019**, *13*, 352–360. [CrossRef]
13. Al-Badi, A. Performance assessment of 20.4 kW eco-house grid-connected PV plant in Oman. *Int. J. Sustain. Eng.* **2020**, *13*, 230–241. [CrossRef]
14. ESMAP (Energy Sector Management Assistance Program). *Global Photovoltaic Power Potential by Country*; World Bank: Washington, DC, USA, 2020; p. 62. Available online: https://documents1.worldbank.org/curated/en/466331592817725242/pdf/Global-Photovoltaic-Power-Potential-by-Country.pdf (accessed on 14 June 2021).
15. IRENA (International Renewable Energy Agency). *Renewables Readiness Assessment: Sultanate of Oman*; International Renewable Energy Agency: Abu Dhabi, United Arab Emirates, 2014; p. 88. Available online: https://www.irena.org/-/media/Files/IRENA/Agency/Publication/2014/IRENA_RRA_Oman_2014_LR.pdf (accessed on 25 September 2021).
16. DATA PORTAL, Labor Market, Data of Year 2018 by National Centre for Statistics & Information (NCSI) of the Sultanate of Oman. Available online: https://data.gov.om/nyavlmf/labor-market (accessed on 23 September 2021).
17. DATA PORTAL, Communication, Data of Year 2018 by National Centre for Statistics & Information (NCSI) of the Sultanate of Oman. Available online: https://data.gov.om/lgtrktf/communication?indicator=No.%20of%20Exchanges (accessed on 23 September 2021).
18. PVGIS Contact Points. Available online: https://ec.europa.eu/jrc/en/PVGIS/about/contact (accessed on 26 September 2021).
19. PVGIS Users Manual. Available online: https://ec.europa.eu/jrc/en/PVGIS/docs/usermanual (accessed on 25 September 2021).
20. CM SAF (Satellite Application Facility on Climate Monitoring). Available online: https://www.cmsaf.eu/EN/Home/home_node.html (accessed on 25 September 2021).
21. PVGIS Data Sources and Calculation Methods. Available online: https://ec.europa.eu/jrc/en/PVGIS/docs/methods (accessed on 25 September 2021).
22. Mueller, R.; Behrendt, T.; Hammer, A.; Kemper, A. A New Algorithm for the Satellite-Based Retrieval of Solar Surface Irradiance in Spectral Bands. *Remote Sens.* **2012**, *4*, 622–647. [CrossRef]
23. Mueller, R.; Matsoukas, C.; Gratzki, A.; Behr, H.D.; Hollmann, R. The CM-SAF operational scheme for the satellite based retrieval of solar surface irradiance—A LUT based eigenvector hybrid approach. *Remote Sens. Environ.* **2009**, *113*, 1012–1024. [CrossRef]
24. Huld, T.; Müller, R.; Gambardella, A. A new solar radiation database for estimating PV performance in Europe and Africa. *Sol. Energy* **2012**, *86*, 1803–1815. [CrossRef]
25. Gracia Amillo, A.; Huld, T.; Müller, R. A New Database of Global and Direct Solar Radiation Using the Eastern Meteosat Satellite, Models and Validation. *Remote Sens.* **2014**, *6*, 8165–8189. [CrossRef]
26. BSRN (Baseline Surface Radiation Network): Baseline Surface Radiation Network. Available online: https://bsrn.awi.de (accessed on 25 September 2021).
27. DATA PORTAL, Population, Data of August 2021 by National Centre for Statistics & Information (NCSI) of the Sultanate of Oman. Available online: https://data.gov.om/amxduh/population (accessed on 23 September 2021).
28. Oman by region—Foreign Ministry of Oman. Available online: https://fm.gov.om/about-oman/state/oman-by-region (accessed on 22 September 2021).
29. About Oman—Omani Ministry of Health. Available online: https://www.moh.gov.om/en/about-oman (accessed on 21 September 2021).

30. Map of the Sultanate of Oman—By the National Survey Authority, Ministry of Defense—Hosted at the Website of the Foreign Ministry of Oman. Available online: https://fm.gov.om/wp-content/uploads/Oman_Map_AR.pdf (accessed on 22 September 2021).
31. Cotfas, D.T.; Cotfas, P.A. Comparative Study of Two Commercial Photovoltaic Panels under Natural Sunlight Conditions. *Int. J. Photoenergy* **2019**, *2019*, 10. [CrossRef]
32. Andreani, L.C.; Bozzola, A.; Kowalczewski, P.; Liscidini, M.; Redorici, L. Silicon solar cells: Toward the efficiency limits. *Adv. Phys. X* **2019**, *4*, 24. [CrossRef]
33. Common Types of Solar Cells—What Are Better Silicon, Monocrystalline, or Polycrystalline Solar Cells? Available online: https://www.altenergy.org/renewables/solar/common-types-of-solar-cells.html (accessed on 26 September 2021).
34. Attari, K.; Elyaakoubi, A.; Asselman, A. Performance analysis and investigation of a grid-connected photovoltaic installation in Morocco. *Energy Rep.* **2016**, *2*, 261–266. [CrossRef]
35. Global Solar Atlas (GSA), data for 23.583889°, 058.407778° (Muscat). Available online: https://globalsolaratlas.info/detail?s=23.583889,58.407778&m=site&c=23.583497,58.40744,11 (accessed on 25 September 2021).
36. Darhmaoui, H.; Lahjouji, D. Latitude Based Model for Tilt Angle Optimization for Solar Collectors in the Mediterranean Region. *Energy Procedia* **2013**, *42*, 26–435. [CrossRef]
37. Dhimish, M.; Silvestre, S. Estimating the impact of azimuth-angle variations on photovoltaic annual energy production. *Clean Energy* **2019**, *3*, 47–58. [CrossRef]
38. Seme, S.; Štumberger, B.; Hadžiselimović, M.; Sredenšek, K. Solar Photovoltaic Tracking Systems for Electricity Generation: A Review. *Energies* **2020**, *13*, 4224. [CrossRef]
39. Hereher, M.; El Kenawy, A.M. Exploring the potential of solar, tidal, and wind energy resources in Oman using an integrated climatic-socioeconomic approach. *Renew. Energy* **2020**, *161*, 662–675. [CrossRef]
40. OPWP (Oman Power and Water Procurement Company). *Solar Data—Weather Impact Analysis*; Oman Power and Water Procurement Company: Muscat, Oman; p. 42. Available online: https://omanpwp.om/PDF/Solar%20Data%20-%20Weather%20Impact%20Analysis.pdf (accessed on 3 November 2021).
41. Introduction to OPWP. Available online: https://omanpwp.om/new/Pages.aspx?Pid=1 (accessed on 6 November 2021).
42. About Nama Group. Available online: https://www.omangrid.com/en/Pages/AMA-Group.aspx (accessed on 6 November 2021).
43. Nama Group. Available online: https://www.linkedin.com/company/namagroupoman (accessed on 6 November 2021).
44. Gulkowski, S.; Zdyb, A.; Dragan, P. Experimental Efficiency Analysis of a Photovoltaic System with Different Module Technologies under Temperate Climate Conditions. *Appl. Sci.* **2019**, *9*, 141. [CrossRef]
45. GeoNames. Available online: https://www.geonames.org/advanced-search.html (accessed on 22 September 2021).
46. Photovoltaic Geographical Information System (PVGIS). Available online: https://re.jrc.ec.europa.eu/pvg_tools/en (accessed on 22 September 2021).
47. Google Maps. Available online: https://www.google.com/maps (accessed on 22 September 2021).

sustainability

MDPI

Article

Exploring the Roles of Renewable Energy, Education Spending, and CO_2 Emissions towards Health Spending in South Asian Countries

Usman Mehmood [1,2,*], Ephraim Bonah Agyekum [3], Salah Kamel [4], Hossein Shahinzadeh [5,6] and Ata Jahangir Moshayedi [7,*]

1 Department of Political Science, University of Management and Technology, Lahore 54770, Pakistan
2 Remote Sensing, GIS and Climatic Research Lab (National Center of GIS and Space Applications), Centre for Remote Sensing, University of the Punjab, Lahore 54590, Pakistan
3 Department of Nuclear and Renewable Energy, Ural Federal University Named after the First President of Russia Boris Yeltsin, 620002 Ekaterinburg, Russia; agyekumephraim@yahoo.com
4 Department of Electrical Engineering, Faculty of Engineering, Aswan University, Aswan 81542, Egypt; skamel@aswu.edu.eg
5 Smart Microgrid Research Center, Najafabad Branch, Islamic Azad University, Najafabad 85141-43131, Iran; h.s.shahinzadeh@ieee.org
6 Department of Electrical Engineering, Amirkabir University of Technology (Tehran Polytechnic), Tehran 15916-34311, Iran
7 School of Information Engineering, Jiangxi University of Science and Technology, Ganzhou 341000, China
* Correspondence: usmanmehmood.umt@gmail.com (U.M.); ajm@jxust.edu.cn (A.J.M.)

Citation: Mehmood, U.; Agyekum, E.B.; Kamel, S.; Shahinzadeh, H.; Moshayedi, A.J. Exploring the Roles of Renewable Energy, Education Spending, and CO_2 Emissions towards Health Spending in South Asian Countries. *Sustainability* **2022**, *14*, 3549. https://doi.org/10.3390/su14063549

Academic Editor: Mostafa Ghasemi Baboli

Received: 7 February 2022
Accepted: 14 March 2022
Published: 17 March 2022

Abstract: This research is mainly aimed at determining the effect of renewable energy (RE), education expenditures, and CO_2 emissions on health expenditures in selected South Asian countries. There is an insufficient number of studies that investigate the linkages between health expenditures (HE) and CO_2 emissions in South Asian countries. This study combined RE and gross domestic product (GDP) to identify their effect on health spending. We utilized the annual data of 1990–2018, and applied FMOLS and DOLS estimators over the panel data of five South Asian countries. According to the DOLS and FMOLS long-run results, GDP, RE, and education expenditures are negatively associated with health expenditures. This suggests that renewable energy puts less pressure on environmental quality, which leads to less health spending in the five South Asian countries studied. The empirical results also show that HE and CO_2 emissions are positively and significantly related, which implies that an increase in CO_2 emissions increases the financial burden on the various countries' health sector. This study, therefore, recommends the usage of renewable sources to improve public health and to help lower health expenditures. To achieve sustainable development, it is also important to increase investment in the educational sector in the various countries.

Keywords: health expenditures; CO_2 emissions; renewable energy; South Asian countries; FMOLS

1. Introduction

Today, breathing in clean air is a blessing, because air is continuously being contaminated by greenhouse gases (GHG) [1–3]. Air pollutants, such as CO_2 and SO_2, create respiratory diseases [4]. Therefore, an increase in air pollution has a negative effect on public health [5,6]. According to the World Health Organization (WHO), about three million people die globally due to air pollution every year [7]. Moreover, an increase in health expenditure (HE) negatively affects labor productivity, which affects economic growth [4]. The social cost of air pollution is of great concern, due to an uncertain estimation of air pollution levels, which, ultimately, affects economic growth [8]. CO_2 is the main pollutant in greenhouse gasses [9,10]. Governments all over the world are, therefore, trying to limit its emissions, due to its harmful effect on public health [4]. Many studies [9,11,12] that have investigated the linkages between HE and GDP have validated that not only does HE boost economic growth, but it also has a significant positive impact on public health.

The need to provide adequate healthcare services has led to rapid development in the health sector across the globe in recent times. There are two views regarding HE. The first view does not support excess HE, such as luxurious commodities [13]. On the other hand, the second point of view believes that HE is compulsory, and, as a result, governmental intervention is necessary [14,15]. Because HE is necessary for the development of a country's economy, such expenditures put pressure on governments [4].

To increase HE, many other socio-economic factors need to be considered, such as population and the percentage of budget allocation to the health sector. Ross and Chia-Ling [16] pointed out that important factors of healthcare include economic growth and level of education. Health conditions can be estimated using HE [17]. Some underdeveloped countries have strong health systems, due to strong economic growth and efficient education systems. Therefore, it is important to probe the factors of HE, which may be environmental (air pollution), economic (GDP), or non-economic factors (education).

In addition to CO_2 emissions, the ecological footprint (EF) is another parameter that takes into consideration the extent of regenerative biological capacity human activities demand from the environment, which is occasioned by the production and consumption of goods and services. EF is defined as *"how much area of biologically productive land and water an individual, population, or activity requires to produce all the resources it consumes and to absorb the waste (carbon dioxide) it generates, using prevailing technology and resource management practices"* [18]. EF is a broad parameter taken for environmental degradation. The aggregation of EF includes grazing land, forest land, cropland, carbon footprint and built-up land [19]. According to Ozturk et al. [20], a positive linkage is found between environmental degradation and EF. The prime determinant of EF is GDP [21], while other determinants are renewable and non-renewable energy, the latter of which is the prime cause of GHG emissions [22–25].

In South Asian countries, the region's vulnerability to many environmental issues has caused a rapid rise in HE [26,27]. HE has also increased considerably in the region as a result of the outbreak of the COVID-19 pandemic [28]. Compared to 2018, HE in 2020increased by 2.11%, 5.55% and 4.28% in India, Pakistan and Sri Lanka, while it decreased by 0.98% and 1.84% in Bangladesh and Nepal, respectively [29]. South Asian countries were faced with a shortage of healthcare supplies, such as ventilators and beds, during the pandemic, due to the region's fragile healthcare systems. These countries had to allocate their development budgets on health expenditures during this period.

A direct relationship is observed between GDP and CO_2 emissions, according to a study by [26]; thus, the negative effect of pollution on health increases [2,3]. Therefore, pollution (i.e., CO_2) and EF are directly related to HE. However, apart from the usage of fossil fuels in industries, renewable sources are also being utilized, which improves air quality [30,31]. Renewable sources have the potential to increase the GDP of a country, without hurting the quality of the environment. The renewable energy sector also holds the capacity to create more jobs to reduce unemployment [32].

Despite the availability of many research studies that investigated the connection between HE and air pollution, there are few studies that probed this association in South Asian countries [14,33,34]. Therefore, the current study fills the research gap by investigating the linkages between HE, renewable energy (RE), and air pollutants in South Asian countries, i.e., Bangladesh, India, Nepal, Pakistan, and Sri Lanka.

This study makes the following contributions: Firstly, this study investigates the association between CO_2, GDP, and HE. Secondly, this research incorporates renewable energy with HE and GDP. Wang et al. [35] explored the linkages among GDP, HE, and CO_2 by applying the autoregressive distributed lag (ARDL) technique. The ARDL approach provides reliable results, without considering sample size, and provides an error correction term (ECT). Khan et al. [36] investigated the linkages between investment, health, and environmental degradation by utilizing canonical co-integration regression (CCP). They employed the VECM and FMOLS panels to assess the linkages. The study shows that HE, air pollution, and renewable energy move together. It also indicates that non-economic factors play a role in health spending.

This paper is organized into five sections. The first section deals with a brief introduction to the social, economic, and environmental factors affecting HE, and the factors responsible for the increasing HE in South Asian countries, followed by a literature review in the second section. The third section focuses on the methodology, while the fourth section provides the results and discussion. Finally, the last section deals with the conclusion and policy implications for sustainable development in South Asian countries.

2. Literature Review

2.1. CO_2 Emissions on Health Expenditures

CO_2 emissions are considered an important factor in the environment, since their increase negatively affects public health. Several studies have investigated the healthcare air pollution nexus [5,12]. Yazdi and Khanalizadeh [37] explored the determinants of HE, and examined MENA countries using data from 1995 to 2014. The ARDL method revealed an association between CO_2, PM10, and HE. They found a positive correlation between the increase in pollutants and HE. The studies of [38,39] also found similar results. Chaabouni et al. [14] examined the linkages between CO_2, HE, and GDP from 1995 to 2013, over the panel of 51 countries. The scholars applied the Gaussian mixture model (GMM) and the dynamic simultaneous equation for the analysis. They found two-way causality between CO_2 and GDP, GDP, and HE. Moreover, one-way causality was found between CO_2 and HE in most economies. According to [40], environmental problems are due to CO_2 emissions that affect public health. They analyzed data from 1990 to 2015 for the African region by applying the ARDL approach. A positive association was found between GDP and HE, but negative linkages were found between CO_2 and HE. Chaabouni and Saidi [41], and Usman et al. [42], also confirmed that air pollution increases the mortality rate.

2.2. Health Expenditures and GDP

HE is increasing rapidly in the world today, and these expenditures are increasing the GDP of different economies. A school of thought proposes that if a country establishes a sound healthcare system, then the life expectancy and social welfare of its people will also increase [42,43]. It, therefore, suggests that an improvement in the health conditions of the people affects their work efficiency positively, which improves a country's economy. Atilgan et al. [44] found that HE improves GDP; if a country spends more on health, it will grow economically. However, the inability of most developing countries to invest in the health sector results in a lower GDP for such economies.

Wang et al. [35] applied the ARDL approach and the Granger causality application to annual data of 1995–2017, and found that HE, CO_2, and GDP move together in the long run. Bidirectional causality was found between HE and CO_2. Omri et al. [45] also conducted research on 13 MENA economies for the annual data of 1995–2005. They found that HE and GDP move together. Few scholars used foreign direct investment (FDI) as a proxy for economic growth, and investigated its association with HE. Usman et al. [42] investigated the association between air pollution, HE, and socio-economic factors. They used CO_2 as a proxy for environmental pollution, FDI and GDP as economic factors, and education and population as non-economic factors. The researchers evaluated the data of 13 countries for 1994–2017. The panel co-integration model found that CO_2, GDP, education, FDI, and HE correlated in the long run. The non-economic factor, i.e., population, was found to be positively associated with public and private HE, while negative linkages were found between education and HE. The studies of [46,47] found a positive association between HE and GDP. Most of the early studies found inconclusive results about the association between HE and GDP. They mostly used GDP as a proxy for economic growth.

2.3. Non-Economic Variables and Health Expenditures

Past studies rarely probed the linkages between non-economic factors and HE; thus, this study attempts to fill this research gap. Yao et al. [47] investigated the association between education and HE. They used the GMM approach over the data of 2001–2016. They found that while education does not affect HE, HE affects education positively. However, [48] found that education puts negative pressures on HE. There are very few studies [42,49] that probed the associations between non-economic factors and HE.

2.4. Health Expenditures and Renewable Energy

Today, the development and use of RE are increasing, in order to meet the world's energy requirements. Khan et al. [36] probed the HE and RE associations with some other factors in ASEAN countries, using 2007–2017 data. They found that RE reduces HE and improves labor production. Therefore, environmental improvement leads to more economic growth. Apergis et al. [50] probed the linkages between healthcare, RE, and GDP in 42 African economies, over the data of 1995–2011. A two-way association was found between RE and CO_2 emissions, and one-way causality was found between RE and healthcare. Apergis and Payne [51], and Alola et al. [52], also made similar conclusions. Furthermore, Jebli [53] found two-way causality between health and RE. Air pollution puts pressure on public health, which negatively affects a country's economic performance. Therefore, RE is a solution for better public health and good economic growth [54].

3. Material and Methods

Annual data of 1995–2019 for 5 South Asian countries were employed for the analysis. All data were obtained from the World Bank Data Indicators (WDI) [29]. HE is taken as current health expenditure, i.e., % of GDP as a dependent variable, GDP as an economic factor, and CO_2 as an air pollutant. RE and education as educational attainment, at least completed lower secondary education, population 25+, total (%) (cumulative) are also independent variables. The general equation of this study is as follows:

$$lnHE_t = \beta_0 + \beta_1 RE_t + \beta_2 G_t + \beta_3 ED_t + \beta_4 CO_t + \varepsilon_t \quad (1)$$

A series of steps were used for the long-run estimations. Initially, we tested the cross-section dependence (CD) within the used data. It provides valuable insights and prevents bias in the outcomes. This study employed the LM test [55] and CD tests [56]. Mathematically, it can be written as follows:

$$LM = T \sum_{i=1}^{n-1} \sum_{j=i+1}^{n} \partial_{ij}^{t} \quad (2)$$

$$CD = \sqrt{\frac{2T}{N(N-1)}} \left(\sum_{i=1}^{n-1} \sum_{j=i+1}^{n} \partial_{ij}^{t} \right) \quad (3)$$

where T and N stand for the period and the number of cross-sections, respectively. Errors' pairwise correlation between i and j is denoted by ∂_{ij}^{t}. After the CD test, in order to test the slope phenomenon, we performed a slope homogeneity test. The homogeneity existence among the panel can cause unreliable and misleading outcomes [57]. The mathematical expressions for slope homogeneity tests of $\tilde{\Delta}$ and $\tilde{\Delta}_{\text{adj}}$ are as follows:are as follows:

$$\tilde{\Delta} = \sqrt{N} \left(\frac{N^{-1}\tilde{S} - K}{\sqrt{2K}} \right) \quad (4)$$

$$\widetilde{\Delta}_{\text{adj}} = \sqrt{N}\left(\frac{N^{-1}\widetilde{S} - E\left(\widetilde{Z}_{iT}\right)}{\sqrt{var\left(\widetilde{Z}_{iT}\right)}}\right) \tag{5}$$

where $var\left(\widetilde{Z}_{iT}\right)$ = 2K (T − K − 1)/(T + 1) and $E\left(\widetilde{Z}_{iT}\right)$ = K. The modified test of $\widetilde{S}$ is as follows:

$$\widetilde{S} = \sum_{i=1}^{n}\left(\widetilde{\gamma}_i - \widetilde{\gamma}_{WFE}\right)\frac{Y_i' M_T X_i}{\widetilde{\varnothing}_i^2}\left(\widetilde{\gamma}_i - \widetilde{\gamma}_{WFE}\right) \tag{6}$$

For an individual unit, the pooled OLS test value is represented by $\widetilde{\gamma}_i$. Similarly, $\widetilde{\gamma}_{WFE}$ and M_T denote the weighted pooled estimator and identity matrix, respectively.

Subsequently, the stationarity among the variables of interest can be checked by unit root tests. The current study utilized two types of unit root tests, i.e., a set of first-generation [58,59] unit root test, as well as second-generation unit root tests. In the case of the order of integration at level I(0), simple regression of ordinary least square can be employed. However, it becomes compulsory to check the cointegration between variables in the case of cointegration at the first difference. Instead of the cointegration test [60–62], the current study incorporated the [63] test that incorporates CD among the data.

After co-integration confirmation, we examined the long-run relationship among variables of interest. To do this, the current study utilized dynamic ordinary least square (DOLS) and fully modified ordinary least square (FMOLS) approaches. Its mathematical expression is as follows:

$$\widetilde{\gamma}_{FMOLS} = \left[N^{-1}\sum_{i=1}^{N}\left(\sum_{t=1}^{t} \cup_{it} - \bar{\cup}_i\right)^2\right]^{-1} \times \left[\left(\sum_{t=1}^{T} \cup_{it} - \bar{\cup}_i\right)\widehat{S}_{it} - T\widehat{\Delta}_{\epsilon\mu}\right] \tag{7}$$

$$\widetilde{\gamma}_{DOLS} = \left[N^{-1}\sum_{i=1}^{N}\left(\sum_{t=1}^{T} C_{it}C_{it}'\right)^{-1}\left(\sum_{t=1}^{T} C_{it}C_{it}'\right)\right] \tag{8}$$

4. Results and Discussion

This section provides the econometric test results. To obtain robust results, this work used widely used methodologies. For this purpose, step-by-step tests of descriptive statistics, a set of unit root tests, a cross-section dependence test, and FMOLS and DOLS tests were applied. The current study implemented the FMOL and DOLS approaches for long-run estimations. Both techniques are capable of eliminating the endogeneity and autocorrelation issues. To uncover the causality among the selected variables, the current study incorporated a test that is efficient enough to eliminate the CS issues in panel data. A discussion on how health expenditure, renewable energy, education expenditures, and GDP are correlated in the long run is conducted. The descriptive statistics are followed by panel unit root tests, panel co-integration, and panel FMLOS and DOLS tests.

Table 1 shows the results of the descriptive statistics, which show that GDP and CO_2 have the highest and the lowest mean values, respectively. GDP has the highest maximum value, while education and CO_2 are negatively skewed. Table 2 shows the empirical outcomes of the unit root tests. It is evident that HE and GDP have a unit root at level I(0). However, HE and GDP are stationary at the first difference. The data of renewable energy, education expenditures, and CO_2 emissions are stationary at the first difference. We, therefore, performed two sets of unit root tests, i.e., [58], [64], and second-generation tests to validate the robustness of the results. The Im, Pesaran, and Shin test results show that GDP, renewable energy, carbon dioxide emissions, and education expenditures have a unit root at level I(0). However, the stationarity of all the other variables is found at the first difference. The CIPS and CADF test results in Table 3 also show similar patterns of findings. Table 4 shows the results of the CD test, with cross-sectional dependence. This is due to the cultural, environmental, and economic dependence among the estimated nations.

Table 1. Descriptive statistics.

	Health Expenditure	Renewable Energy	GDP	Education	CO_2
Mean	1.263758	4.079114	6.755095	1.029647	−0.864372
Median	1.217133	4.053435	6.684298	1.060974	−0.541043
Maximum	1.906449	4.555136	8.200345	1.544385	0.548122
Minimum	0.811057	3.598303	5.878314	0.354242	−3.386252
Std. Dev.	0.281230	0.268093	0.568801	0.307137	0.872248
Skewness	0.698690	0.250900	0.635160	−0.124949	−0.576190
Kurtosis	2.522287	2.092878	2.714132	2.046960	2.419577

Table 2. LLC and IPS unit root tests.

Variable	Im, Pesaran, Shin		Levin-Lin-Chu Unit Root Test	
	Level	1st Difference	Level	1st Difference
$lnHE_t$	−1.47 *	−5.34 ***	−1.24	−6.85 ***
lnG_t	6.44	−3.18 ***	4.16	−2.61 **
$lnRE_t$	0.26	−2.99 **	−1.34**	−0.97
$lnED_t$	−0.63	−6.47 ***	−1.70**	−5.11 ***
$lnCO_{2t}$	0.39	−4.04 ***	−1.57**	−2.12 **

***, ** and * represent significant levels of 1% and 5%, respectively.

Table 3. CIPS and CADF unit root tests.

Variable	CIPS test		CADF test	
	Level	First Difference	Level	First Difference
$lnCO_{2t}$	0.47	−2.53 **	−1.45	−4.22 **
$lnHE_t$	−2.50	−5.40 ***	−1.87	−2.53 **
$lnRE_t$	−3.62 **	−5.40 ***	−3.51 ***	−4.84 ***
lnG_t	−3.31 ***	−5.72 ***	−2.76	−4.52 ***
$lnED_t$	−1.41	−2.46 **	−3.12 **	−2.35 *

***, ** and * represent significant levels of 1%, 5% and 10%, respectively.

Table 4. Cross-section dependence test.

	lnHE	lnG	lnRE	lnED	$lnCO_2$
CD test	17.02 *** (0.00)	17.04 *** (0.00)	17.01 *** (0.00)	17.02 *** (0.00)	15.81 *** (0.00)
LM test	289.91 *** (0.00)	289.88 *** (0.00)	290.84 *** (0.00)	289.91 *** (0.00)	251.04 *** (0.00)

*** represents significant level of 1%.

Table 5 is the Westerlund [63] test results, which confirm the long-run association among the estimated variables. After confirming the co-integration, the long- and short-run analyses can then be conducted. According to the DOLS and FMOLS long-run results, GDP, RE, and education expenditures are negatively associated with health expenditures, as shown in Tables 6 and 7. These results are in line with [4]. These results imply that investing in the renewable energy sector improves air quality, which puts less pressure on health expenditures. At the same time, when the government spends more money on education, people tend to pay more attention to their health and improve their health status, which reduces a country's health expenditure significantly. GDP is also negatively associated with health expenditure, which means that a rise in GDP reduces health expenditure in South Asian countries. These results suggest that these countries are experiencing economic growth by compromising on the health of the people. Their spending on education is not high enough to improve the health of the people. The governments of these five countries must increase their health expenditure as economic growth increases. CO_2 emissions are positively related to HE. This is an indication that a rise in CO_2 emissions will also increase health expenditure in South Asian countries. This finding is consistent with that reported in [4].

Table 5. Westerlund test.

Stats	G_t	G_a	P_t	P_a
value	−1.22	−0.61 ***	−1.7	−0.78 ***
Z-value	4.04	4.33	4.17	3.40
Prob	1.00	1.00	1.00	1.00
Robust prob	0.80	0.00	0.55	0.00

*** represents significant level of 1%.

Table 6. FMOLS and DOLS test results with CO_2.

Variables	FMOLS	Prob	DOLS	Prob
lnRE	−0.41 ***	0.00	−0.49 ***	0.03
lnG	−0.86 ***	0.00	−0.78 ***	0.00
lnED	−0.08 ***	0.00	−0.07 ***	0.00
lnCO_2	0.96 ***	0.00	0.96 ***	0.00

*** represents significant level of 1%.

Table 7. FMOLS and DOLS test results with EF.

Variables	FMOLS	Prob	DOLS	Prob
lnRE	−0.41 ***	0.00	−0.97 ***	0.03
lnG	−0.19 ***	0.01	−1.69 ***	0.01
lnED	−0.07 ***	0.05	−0.84 ***	0.05
lnEF	0.21 ***	0.00	0.67 ***	0.00

*** represents significant level of 1%.

5. Conclusion and Recommendations

It has been widely discussed that both developed and developing countries are facing the problems of drastic environmental changes. Considering the vulnerabilities of South Asian countries, this work attempts to highlight some important factors of health expenditure. Therefore, this study mainly aims to find associations between health expenditure, CO_2 emissions, education expenditure, and renewable energy in selected South Asian countries. This work utilized the annual data of 1990–2018, and applied FMOLS and DOLS estimators over the panel data of five South Asian countries. The panel unit root tests were employed to check the stationarity of the data. After the order of integration confirmation in time series, we applied the Westerlund co-integration test to determine the integration between the time series. FMOLS and DOLS tests were applied to check the association among the data, to show how the dependent and independent variables are inter-related.

The empirical results show that HE and CO_2 emissions are positively and significantly related, which implies that an increase in CO_2 emissions increases the financial burden on the health sector in the various countries. Moreover, the FMOLS and DOLS tests revealed that the use of renewable energy reduces health expenditure in the five South Asian countries studied. GDP is also negatively associated with health expenditure, which means that a rise in GDP reduces health expenditure in the five South Asian countries. Renewable energy puts less pressure on environmental quality, which leads to less spending on health. The policy implication of this study is that South Asian countries need to invest in clean energy sources, such as geothermal, solar, wind, waste-to-energy, etc. It is also important for the countries to implement measures that promote efficiency in their energy sector. In doing so, these governments should consider the implementation of financial policies, such as tax waivers, and the provision of incentives to individuals and companies to use RE, since the initial cost of RE technologies is generally high. Subsidies on fossil fuel products can also be removed if such a policy exists in any of the countries, and the proceeds would be channeled into RE development. Various industries should be encouraged to use efficient and advanced technologies during production, to help cut down on their emissions and enhance economic growth. Taxes can, therefore, be imposed on firms whose activities result in heavy air pollution.

The various countries must also increase their investment into their education sector, particularly in the area of innovation in the RE sector. These countries should also consider amending aspects of their educational curriculum to include energy efficiency, environmental protection, and RE studies beginning from the basic level of education. Educating the populace is one of the possible ways that government policies on the environment can be institutionalized. These steps will ultimately lower the health expenditure of these developing countries.

Apart from the contribution of this work, future research can analyze the data for some gulf and other groups of countries to form effective environmental policies. Future work can use country-specific data by adopting the econometric techniques in this study.

Author Contributions: Conceptualization, U.M.; Data curation, U.M., E.B.A.; Formal analysis, U.M., E.B.A., S.K., H.S. and A.J.M.; Funding acquisition, E.B.A., S.K., H.S. and A.J.M.; Investigation, U.M., E.B.A., S.K., H.S. and A.J.M.; Methodology, U.M.; Project administration, U.M.; Software, U.M.; Validation, U.M., E.B.A., S.K., H.S. and A.J.M.; Writing—original draft, U.M., E.B.A., S.K., H.S. and A.J.M.; Writing—review and editing, U.M., E.B.A., S.K., H.S. and A.J.M. All authors have read and agreed to the published version of the manuscript.

Funding: This work was supported by Jiangxi University of Science and Technology, 341000, Ganzhou, P.R China, under funding number: 2021205200100563.

Institutional Review Board Statement: Not applicable.

Informed Consent Statement: Not applicable.

Data Availability Statement: Data used, and their sources are provided in the text.

Conflicts of Interest: The authors declare no conflict of interest.

References

1. Mehmood, U. Globalization-Driven CO_2 Emissions in Singapore: An Application of ARDL Approach. *Environ. Sci. Pollut. Res.* **2020**, *28*, 11317–11322. [CrossRef] [PubMed]
2. Qayyum, F.; Mehmood, U.; Tariq, S.; Nawaz, H. Particulate Matter (PM2. 5) and Diseases: An Autoregressive Distributed Lag (ARDL) Technique. *Environ. Sci. Pollut. Res.* **2021**, *28*, 67511–67518. [CrossRef] [PubMed]
3. Mehmood, U.; Azhar, A.; Qayyum, F.; Nawaz, H.; Tariq, S. Air Pollution and Hospitalization in Megacities: Empirical Evidence from Pakistan. *Environ. Sci. Pollut. Res.* **2021**, *28*, 51384–51390. [CrossRef] [PubMed]
4. Mujtaba, G.; Shahzad, S.J.H. Air Pollutants, Economic Growth and Public Health: Implications for Sustainable Development in OECD Countries. *Environ. Sci. Pollut. Res.* **2020**, *28*, 12686–12698. [CrossRef]
5. Li, Y.; Chiu, Y.; Lin, T.-Y. The Impact of Economic Growth and Air Pollution on Public Health in 31 Chinese Cities. *Int. J. Environ. Res. Public Health* **2019**, *16*, 393. [CrossRef]

6. Kayani, G.M.; Ashfaq, S.; Siddique, A. Assessment of Financial Development on Environmental Effect: Implications for Sustainable Development. *J. Clean. Prod.* **2020**, *261*, 120984. [CrossRef]
7. WHO. WHO Releases Country Estimates on Air Pollution Exposure and Health Impact. Available online: https://www.who.int/news/item/27-09-2016-who-releases-country-estimates-on-air-pollution-exposure-and-health-impact (accessed on 24 February 2022).
8. Lu, Z.-N.; Chen, H.; Hao, Y.; Wang, J.; Song, X.; Mok, T.M. The Dynamic Relationship between Environmental Pollution, Economic Development and Public Health: Evidence from China. *J. Clean. Prod.* **2017**, *166*, 134–147. [CrossRef]
9. Zaidi, S.; Saidi, K. Environmental Pollution, Health Expenditure and Economic Growth in the Sub-Saharan Africa Countries: Panel ARDL Approach. *Sustain. Cities Soc.* **2018**, *41*, 833–840. [CrossRef]
10. Apergis, N.; Garzón, A.J. Greenhouse Gas Emissions Convergence in Spain: Evidence from the Club Clustering Approach. *Environ. Sci. Pollut. Res.* **2020**, *27*, 38602–38606. [CrossRef]
11. Mărginean, I. Public Expenditure with Education and Healthcare in EU Countries. *Procedia Econ. Financ.* **2014**, *8*, 429–435. [CrossRef]
12. Murthy, V.N.R.; Okunade, A.A. Determinants of U.S. Health Expenditure: Evidence from Autoregressive Distributed Lag (ARDL) Approach to Cointegration. *Econ. Model.* **2016**, *59*, 67–73. [CrossRef]
13. Azam, M.; Khan, A.Q.; Ozturk, I. The Effects of Energy on Investment, Human Health, Environment and Economic Growth: Empirical Evidence from China. *Environ. Sci. Pollut. Res.* **2019**, *26*, 10816–10825. [CrossRef] [PubMed]
14. Chaabouni, S.; Zghidi, N.; Ben Mbarek, M. On the Causal Dynamics between CO_2 Emissions, Health Expenditures and Economic Growth. *Sustain. Cities Soc.* **2016**, *22*, 184–191. [CrossRef]
15. Ye, L.; Zhang, X. Nonlinear Granger Causality between Health Care Expenditure and Economic Growth in the OECD and Major Developing Countries. *Int. J. Environ. Res. Public Health* **2018**, *15*, 1953. [CrossRef]
16. Ross, C.E.; Wu, C.-L. The Links between Education and Health. *Am. Sociol. Rev.* **1995**, *60*, 719–745. [CrossRef]
17. Mirmirani, S.; Li, H.C.; Ilacqua, J.A. Health Care Efficiency in Transition Economies: An Application of Data Envelopment Analysis. *Int. Bus. Econ. Res. J. (IBER)* **2011**, *7*. [CrossRef]
18. Rudolph, A.; Figge, L. Determinants of Ecological Footprints: What Is the Role of Globalization? *Ecol. Indic.* **2017**, *81*, 348–361. [CrossRef]
19. Baloch, M.A.; Zhang, J.; Iqbal, K.; Iqbal, Z. The Effect of Financial Development on Ecological Footprint in BRI Countries: Evidence from Panel Data Estimation. *Environ. Sci. Pollut. Res.* **2019**, *26*, 6199–6208. [CrossRef]
20. Ozturk, I.; Al-Mulali, U.; Saboori, B. Investigating the Environmental Kuznets Curve Hypothesis: The Role of Tourism and Ecological Footprint. *Environ. Sci. Pollut. Res.* **2016**, *23*, 1916–1928. [CrossRef]
21. Anser, M.K.; Apergis, N.; Syed, Q.R.; Alola, A.A. Exploring a New Perspective of Sustainable Development Drive through Environmental Phillips Curve in the Case of the BRICST Countries. *Environ. Sci. Pollut. Res.* **2021**, *28*, 48112–48122. [CrossRef]
22. Sharif, A.; Godil, D.I.; Xu, B.; Sinha, A.; Khan, S.A.R.; Jermsittiparsert, K. Revisiting the Role of Tourism and Globalization in Environmental Degradation in China: Fresh Insights from the Quantile ARDL Approach. *J. Clean. Prod.* **2020**, *272*, 122906. [CrossRef]
23. Nathaniel, S.P. Ecological Footprint, Energy Use, Trade, and Urbanization Linkage in Indonesia. *GeoJournal* **2021**, *86*, 2057–2070. [CrossRef]
24. Adebayo, T.S.; Agyekum, E.B.; Kamel, S.; Zawbaa, H.M.; Altuntaş, M. Drivers of Environmental Degradation in Turkey: Designing an SDG Framework through Advanced Quantile Approaches. *Energy Rep.* **2022**, *8*, 2008–2021. [CrossRef]
25. Zhao, Y.; Ramzan, M.; Adebayo, T.S.; Oladipupo, S.D.; Adeshola, I.; Agyekum, E.B. Role of Renewable Energy Consumption and Technological Innovation to Achieve Carbon Neutrality in Spain: Fresh Insights from Wavelet Coherence and Spectral Causality Approaches. *Front. Environ. Sci.* **2021**, *9*, 769067. [CrossRef]
26. Tariq, S. CO_2 Emissions from Pakistan and India and their relationship with economic variables. *Appl. Ecol. Environ. Res.* **2017**, *15*, 1301–1312. [CrossRef]
27. Mehmood, U.; Tariq, S. Globalization and CO_2 Emissions Nexus: Evidence from the EKC Hypothesis in South Asian Countries. *Environ. Sci. Pollut. Res.* **2020**, *27*, 37044–37056. [CrossRef]
28. Apergis, N.; Bhattacharya, M.; Hadhri, W. Health Care Expenditure and Environmental Pollution: A Cross-Country Comparison across Different Income Groups. *Environ. Sci. Pollut. Res.* **2020**, *27*, 8142–8156. [CrossRef]
29. WDI. *World Development Indicators 2019*; The World Bank: Washington, DC, USA, 2019.
30. Apergis, N.; Payne, J.E.; Apergis, N.; Payne, J. Renewable Energy Consumption and Growth in Eurasia. *Energy Econ.* **2010**, *32*, 1392–1397. [CrossRef]
31. Fernando, Y.; Bee, P.S.; Jabbour, C.J.C.; Thomé, A.M.T. Understanding the Effects of Energy Management Practices on Renewable Energy Supply Chains: Implications for Energy Policy in Emerging Economies. *Energy Policy* **2018**, *118*, 418–428. [CrossRef]
32. Liu, M.; Ren, X.; Cheng, C.; Wang, Z. The Role of Globalization in CO_2 Emissions: A Semi-Parametric Panel Data Analysis for G7. *Sci. Total Environ.* **2020**, *718*, 137379. [CrossRef]
33. Gündüz, M. Healthcare Expenditure and Carbon Footprint in the USA: Evidence from Hidden Cointegration Approach. *Eur. J. Health Econ.* **2020**, *21*, 801–811. [CrossRef]
34. Lu, X.; Yao, T.; Fung, J.C.; Lin, C. Estimation of Health and Economic Costs of Air Pollution over the Pearl River Delta Region in China. *Sci. Total Environ.* **2016**, *566*, 134–143. [CrossRef] [PubMed]

35. Wang, Z.; Asghar, M.M.; Zaidi, S.A.H.; Wang, B. Dynamic Linkages among CO 2 Emissions, Health Expenditures, and Economic Growth: Empirical Evidence from Pakistan. *Environ. Sci. Pollut. Res.* **2019**, *26*, 15285–15299. [CrossRef] [PubMed]
36. Khan, M.K.; Teng, J.Z.; Khan, M.I.; Khan, M.O. Impact of Globalization, Economic Factors and Energy Consumption on CO_2 Emissions in Pakistan. *Sci. Total Environ.* **2019**, *688*, 424–436. [CrossRef] [PubMed]
37. Yazdi, S.K.; Khanalizadeh, B. Air Pollution, Economic Growth and Health Care Expenditure. *Econ. Res.* **2017**, *30*, 1181–1190. [CrossRef]
38. Tran, H.; Kim, J.; Kim, D.; Choi, M.; Choi, M. Impact of Air Pollution on Cause-Specific Mortality in Korea: Results from Bayesian Model Averaging and Principle Component Regression Approaches. *Sci. Total Environ.* **2018**, *636*, 1020–1031. [CrossRef]
39. Wang, H.; Zhang, X. Spatial Heterogeneity of Factors Influencing Transportation CO_2 Emissions in Chinese Cities: Based on Geographically Weighted Regression Model. *Air Qual. Atmos. Health* **2020**, *13*, 977–989. [CrossRef]
40. Saidi, K.; Hammami, S. The Impact of CO_2 Emissions and Economic Growth on Energy Consumption in 58 Countries. *Energy Rep.* **2015**, *1*, 62–70. [CrossRef]
41. Chaabouni, S.; Saidi, K. The Dynamic Links between Carbon Dioxide (CO_2) Emissions, Health Spending and GDP Growth: A Case Study for 51 Countries. *Environ. Res.* **2017**, *158*, 137–144. [CrossRef]
42. Usman, M.; Ma, Z.; Zafar, M.W.; Haseeb, A.; Ashraf, R.U. Are Air Pollution, Economic and Non-Economic Factors Associated with Per Capita Health Expenditures? Evidence from Emerging Economies. *Int. J. Environ. Res. Public Health* **2019**, *16*, 1967. [CrossRef]
43. Omri, A.; Nguyen, D.K. On the Determinants of Renewable Energy Consumption: International Evidence. *Energy* **2014**, *72*, 554–560. [CrossRef]
44. Atilgan, E.; Kilic, D.; Ertugrul, H.M. The Dynamic Relationship between Health Expenditure and Economic Growth: Is the Health-Led Growth Hypothesis Valid for Turkey? *Eur. J. Health Econ.* **2016**, *18*, 567–574. [CrossRef] [PubMed]
45. Omri, A.; Nguyen, D.K.; Rault, C. Causal Interactions between CO_2 Emissions, FDI, and Economic Growth: Evidence from Dynamic Simultaneous-Equation Models. *Econ. Model.* **2014**, *42*, 382–389. [CrossRef]
46. Mehrara, M.; Rezaei, S.; Razi, D.H. Determinants of Renewable Energy Consumption among ECO Countries; Based on Bayesian Model Averaging and Weighted-Average Least Square. *Int. Lett. Soc. Humanist. Sci.* **2015**, *54*, 96–109. [CrossRef]
47. Yao, W.; Gao, D.; Sheng, P. The Impact of Education on Healthcare Expenditure in China: Quantity or Quality. *Appl. Econ. Lett.* **2018**, *26*, 1192–1195. [CrossRef]
48. Kais, S.; Sami, H. An Econometric Study of the Impact of Economic Growth and Energy Use on Carbon Emissions: Panel Data Evidence from Fifty Eight Countries. *Renew. Sustain. Energy Rev.* **2016**, *59*, 1101–1110. [CrossRef]
49. Wang, K.-M.; Lee, Y.-M.; Lin, C.-L.; Tsai, Y.-C. The Effects of Health Shocks on Life Insurance Consumption, Economic Growth, and Health Expenditure: A Dynamic Time and Space Analysis. *Sustain. Cities Soc.* **2018**, *37*, 34–56. [CrossRef]
50. Apergis, N.; Ben Jebli, M.; Ben Youssef, S. Does Renewable Energy Consumption and Health Expenditures Decrease Carbon Dioxide Emissions? Evidence for Sub-Saharan Africa Countries. *Renew. Energy* **2018**, *127*, 1011–1016. [CrossRef]
51. Apergis, N.; Payne, J.E. CO_2 Emissions, Energy Usage, and Output in Central America. *Energy Policy* **2009**, *37*, 3282–3286. [CrossRef]
52. Alola, A.A.; Bekun, F.V.; Sarkodie, S.A. Dynamic Impact of Trade Policy, Economic Growth, Fertility Rate, Renewable and Non-Renewable Energy Consumption on Ecological Footprint in Europe. *Sci. Total Environ.* **2019**, *685*, 702–709. [CrossRef]
53. Ben Jebli, M. On the Causal Links between Health Indicator, Output, Combustible Renewables and Waste Consumption, Rail Transport, and CO_2 Emissions: The Case of Tunisia. *Environ. Sci. Pollut. Res.* **2016**, *23*, 16699–16715. [CrossRef] [PubMed]
54. Buonocore, J.J.; Luckow, P.; Norris, G.; Spengler, J.D.; Biewald, B.; Fisher, J.; Levy, J.I. Health and Climate Benefits of Different Energy-Efficiency and Renewable Energy Choices. *Nat. Clim. Change* **2016**, *6*, 100–105. [CrossRef]
55. Breusch, T.S.; Pagan, A.R. The Lagrange Multiplier Test and Its Applications to Model Specification in Econometrics. *Rev. Econ. Stud.* **1980**, *47*, 239–253. [CrossRef]
56. Pesaran, M.H. General Diagnostic Tests for Cross Section Dependence in Panels. In *Cambridge Working Papers in Economics*; Faculty of Economics: Cambridge, UK, 2004.
57. Le, H.P.; Ozturk, I. The Impacts of Globalization, Financial Development, Government Expenditures, and Institutional Quality on CO_2 Emissions in the Presence of Environmental Kuznets Curve. *Environ. Sci. Pollut. Res.* **2020**, *27*, 22680–22697. [CrossRef]
58. Levin, A.; Lin, C.-F.; James Chu, C.-S. Unit Root Tests in Panel Data: Asymptotic and Finite-Sample Properties. *J. Econom.* **2002**, *108*, 1–24. [CrossRef]
59. Pesaran, M.H.; Shin, Y.; Smith, R.J. Bounds Testing Approaches to the Analysis of Level Relationships. *J. Appl. Econom.* **2001**, *16*, 289–326. [CrossRef]
60. Johansen, S.; Juselius, K. Maximum Likelihood Estimation and Inference on Co-Integration. *Oxf. Bull. Econ. Stat.* **1990**, *52*, 196–210.
61. Kao, C. Spurious Regression and Residual-Based Tests for Cointegration in Panel Data. *J. Econom.* **1999**, *90*, 1–44. [CrossRef]
62. Pedroni, P. Critical Values for Cointegration Tests in Heterogeneous Panels with Multiple Regressors. *Oxf. Bull. Econ. Stat.* **1999**, *61*, 653–670. [CrossRef]
63. Westerlund, J.; Edgerton, D.L. A Panel Bootstrap Cointegration Test. *Econ. Lett.* **2007**, *97*, 185–190. [CrossRef]
64. Im, K.S.; Pesaran, M.H.; Shin, Y. Testing for Unit Roots in Heterogeneous Panels. *J. Econom.* **2003**, *115*, 53–74. [CrossRef]

MDPI
St. Alban-Anlage 66
4052 Basel
Switzerland
Tel. +41 61 683 77 34
Fax +41 61 302 89 18
www.mdpi.com

Sustainability Editorial Office
E-mail: sustainability@mdpi.com
www.mdpi.com/journal/sustainability

www.ingramcontent.com/pod-product-compliance
Ingram Content Group UK Ltd.
Pitfield, Milton Keynes, MK11 3LW, UK
UKHW060124010826
14089UKWH00006B/149

9 783036 539416